A BROKEN TREE

A BROKEN TREE

How DNA Exposed a Family's Secret

Stephen F. Anderson

ROWMAN & LITTLEFIELD
Lanham • Boulder • New York • London

Published by Rowman & Littlefield
An imprint of The Rowman & Littlefield Publishing Group, Inc.
4501 Forbes Boulevard, Suite 200, Lanham, Maryland 20706
www.rowman.com

86-90 Paul Street, London EC2A 4NE

British Library Cataloguing in Publication Information Available

Library of Congress Cataloging-in-Publication Data

Names: Anderson, Stephen F. (Genealogist), author.
Title: A broken tree : how DNA exposed a family's deepest secrets / Stephen F. Anderson.
Description: Lanham: Rowman & Littlefield, [2019]. Includes bibliographical references and index.
Identifiers: LCCN 2019019761 (cloth) | LCCN 2019980163 (ebook) | ISBN 9781538127421 (cloth) | ISBN 9781666909746 (ebook) | ISBN 9781538185414 (paper)
Subjects: Anderson, Stephen F. (Genealogist)--Family. |Anderson family. | Genealogists—United States—Biography. | Illegitimate children—United States—Biography. | Genetic genealogy—United States—Case studies. | Family secrets—United States.
Classification: LCC CT275.A7283 | A3 2019 LCC CT275.A7283 (ebook) | DDC 929.20973
LC record available at https://lccn.loc.gov/2019019761
LC ebook record available at https://lccn.loc.gov/2019980163

CONTENTS

ACKNOWLEDGMENTS

An undertaking of this kind is no small effort. Without the help of several people, I never would have learned the truth about my family's story.

I would first like to thank my brother Tim. I have to say that if it wasn't for Tim, this book may never have been written.

As with all great adventures, you really need a traveling companion to make the adventure complete. Tim was that traveling companion. We bounced ideas back and forth constantly. Early on, I discovered that my way of dealing with every new bit of information we discovered was to talk about it, and talk I did. Tim patiently listened to me and offered excellent feedback to help keep me on track, and to keep me from driving everyone else crazy. Thanks for everything, Tim!

Of course, I want to thank all of my brothers and sisters. Without them, there would never have been a story. They were supportive and willing to step into the unknown when it came time to have their DNA tested, not knowing what their DNA tests would reveal. Best of all, they showed the patience of Job when it came to tolerating my constant questions and need to talk things through. I am eternally grateful that we are family.

I want to give a special thanks to my dear wife and my four children. My wife was always there to keep me grounded, continually reminding me that regardless of what I found in respect to my DNA, I was still the man she married, before any of this happened. She constantly assured me that she wouldn't change a thing. My children loved hearing the constantly evolving stories that came with each new DNA test and interview. They all had a vested interest in my research, because my bloodline is their bloodline. They simply wanted to know the truth, and seemed to enjoy the journey. I appreciate my family's loving support and their continued encouragement to see this through to the end.

A special thanks goes to Jack Anderson of Andergene Labs. Without Jack, I doubt I would have found the answers to many of my questions. I appreciate the many recommendations he provided, and for his interest in seeing each new chapter of my family's story play out. Jack, I think the story is finally over—at least for now. I'm sure there will eventually be more to come.

I also want to thank the folks at AncestryDNA. Where Andergene Labs helped us discover who we were *not* related to, AncestryDNA and their remarkable database of information helped us discover all the people we really were related to. The folks at AncestryDNA really are bringing families together. I am extremely grateful for all their help.

I am also very grateful for my agent, Veronica Park. When she got involved, things started happening, in a very good way. She knows her stuff.

Thanks also to the good folks at Rowman & Littlefield. They made publishing this story so much easier than I ever dreamed possible. You folks are great. Thank you to Suzanne, Mary, and everyone else who helped to bring this book to fruition.

Section 1

The Stories

I

MY FAMILY

When I was a young boy, I loved watching *Leave It to Beaver*. Every day after school, I hurried home just so I could watch it before my dad turned on the evening news. I was fascinated by this show. June Cleaver was the perfect mother. I had never seen such perfection in a mother before—not in my mom, or any of my friends' moms. Oh, how I wished I had June Cleaver for my mother. Her home was spotless in every way. Each night, Ward and the boys came home to a home-cooked meal, complete with dessert and a beautiful smile on June's face. It's funny how these things meant so much to a little boy like me. Ward Cleaver was the perfect father. I don't remember him ever raising his voice to his sons, to the level that anyone could possibly consider an honest-to-goodness yell. You could always count on Wally or the Beaver to get into trouble in every episode, but it was nothing that would merit a good grounding or revocation of privileges, much less a night in the local jail. And where were the sisters? I honestly wondered if that was one of the reasons June and Ward Cleaver seemed so calm most of the time. I had five sisters, so I knew their potential for causing excitement in a family.

I was fascinated by this TV show, yet as entertaining as it was to watch this flawless family living out their lives, I knew this kind of life was far beyond what my own family could ever realistically hope to achieve. I didn't realize at the time that this model was rarely, if ever, truly achieved by any family in America, much less the rest of the world.

My family was about as far away from the Cleavers as anyone could imagine. Our parents were Linda and Mark Anderson. There were nine children in our family, five girls and four boys. I was number seven in the lineup. Starting with the oldest, we had three girls, a boy, a girl, followed by three more boys, and finishing up with a girl. It seemed like a pretty good arrangement, as far as I could tell.

As it turned out, my mom didn't care much for kids. I couldn't for the life of me figure out why she'd had so many children if she didn't like taking care of them, but that's just how it was in our home. Because of that, the first three girls pretty much raised the last three kids. The middle three were more or less left to take care of themselves. Don't get me wrong: My parents never outright abandoned us, nor did they leave all the discipline to the older sisters. My parents were there for the most part, and when necessary, they would bring discipline and order to the chaos of our home life.

My father worked for a local manufacturing company as a salesman, selling fire trucks for a living. This always seemed a bit funny to me. I remember the Jewel Tea man who came by to sell his hardware products, and the vacuum cleaner salesman who was always trying to get us to buy a new Hoover. But as a young boy, I couldn't imagine someone knocking door to door, trying to get someone to buy a fire truck. It simply never made sense to me. It wasn't until I was older that it finally dawned on me that somebody had to be out there hustling up the city officials who were responsible for buying fire trucks for their community's fire department. To this day, I have never met a fire truck salesman other than my dad.

Actually, his job was kind of cool; when we had Career Day at school and we had to report on what our parents did, I never had to worry about duplicating any other kids' accounts. I knew I had something very unique. When Dad sold a fire truck, he always delivered it personally to the town he sold it to. It was unimaginable to him to send someone else to deliver the final product. He wanted to be the only face that a fire department captain or a city official would associate with their purchase of a brand-new truck. He took great pride in what he did, and knew he could guarantee with complete confidence each truck he delivered.

The night before he delivered the truck, he would bring it home so he could leave bright and early the next morning to make the delivery. From time to time, he would get started a little later than usual. When that happened, he would load us younger kids into the fire truck and personally deliver us to school. Just before he got to the corner of the elementary school playground, Dad would turn on the siren and ring the bells. It was like a full-scale alarm. You could hear that siren and those bells throughout the entire town. At our school, all the kids had to play outside and wait until the school bell rang before they were allowed into the building to start the day. Dad made sure we got there a bit early so that every kid, from kindergarten through the sixth grade, would witness our grand arrival. With the sirens still sounding, Tim, Carlee, and I would step down from the brand-new fire truck and make our way through the crowd of kids. It was such a wonderful experience to be given so much attention by our dad. He wanted to make sure our schoolmates remembered us with flair, and parental attention. We felt like we were special, and all the kids at school let us know how lucky we were to have a dad who would do this for us.

As a salesman, Dad spent the work week traveling through several states, meeting with community fire chiefs and city council members in an effort to convince them to spend a lot of money on one very big truck. Because of this, Dad wasn't home much. When he was home, I have no doubt that with nine kids, he had a lot of damage control to address. I am sure this task alone kept him very busy. Dad was also a World War II vet, so he spent a lot of time drinking with his war buddies at the local Veterans of Foreign Wars (VFW) club, a hot spot on weekends. Dad was right in the middle of all VFW affairs.

As far as I can remember, Mom worked at several different jobs during my childhood. She came from an affluent family, so she wasn't too interested in settling down to a life of housekeeping and raising kids, especially since she was on her own five days a week when Dad was on the road.

Maybe this is why June Cleaver fascinated me so much. She genuinely seemed to love being at home, taking care of the place and making sure her boys were well cared for. On top of that, she did it in a perfectly ironed dress that never showed a spot of dirt or food. She was remarkable. I couldn't understand why she would enjoy it so. No woman I knew ever felt that way about keeping house. Add nine kids to that equation, and the thought of enjoying being a stay-at-home mom seemed utterly unrealistic. Such was the case with my mother.

Mom often left the older three girls to care for the youngest three, which included me. On average, there was about a ten-year difference in ages between the older girls and the three youngest. This means that when I was four years old, my sister Gloria, who was responsible for my care, was fourteen. Decades later, we three youngest still have a special bond with our older sisters because of the care they provided us. They weren't perfect caregivers, but they made sure that none of us died, and as far as I can remember, our diapers were changed and we were fed. They were more mothers to us than our own actual mother was. Unfortunately, the middle three didn't fare nearly so well. They were left alone much of the time to fend for themselves. One thing I've learned in life is that young kids left alone don't usually make the best decisions; in fact, they often make some seriously bad ones.

As I was growing up, I remember hearing my oldest sister, Holly, talk about how different the family was before the three youngest were born—even something as basic as what kind of food was on hand. She told me that the older kids never had any fruit to eat when there were six kids in the family. It wasn't until the older ones had moved out and the number of kids in the house dropped to a more manageable size that Mom started buying fruit when she went grocery shopping. I was surprised to hear that; I had always assumed that the fruit available to us younger kids—just the basics, like apples and oranges—had also been available to the older ones. It's not like we're talking about mangos, kiwis, or other exotic fruits. I guess I was wrong. I was left to wonder what the early years were really like.

Of course, when I heard this story, I had to find out more.

When I was in my mid-forties, I was working with my town's historical society, gathering dozens of oral histories from many of the original settlers. I felt comfortable talking with people and capturing their stories on my tape recorder. It was during this time that I decided I was going to fly back to my hometown and visit my oldest sister, Holly, who still lived there. I brought my tape recorder, ready to spend as many hours as it took to capture all the memories she had stored away in her head. When I asked Holly if I could record her stories of our early family, she made it clear that this was not going to happen. She was not going to share a single recollection with me.

I couldn't believe what she was saying. What could have happened during those years that would make my own sister so unwilling to share what was going on in our family? By now I was really intrigued. I knew there was something there that we had to get on tape, so I started pushing the issue. I wouldn't let her brush it off. After barraging her with a list of reasons why she needed to let me record her memories, she simply said, "Wait until Mom and Dad are both gone, and then come back with plenty of tapes, because I have a lot to tell you." Although I begged and pleaded with her, it was pointless. When Holly said she wasn't going to do something, you knew it wasn't going to happen until hell froze over. Even then, it would only be on her terms.

After Dad died, I approached Holly about the interview, but she said the time still wasn't right; we had to wait until Mom was gone, too. Unfortunately, Holly developed cancer and died a few years before our mom died. When Holly was on her deathbed, my sister Gloria talked to her about passing on the stories she had planned to share with me after our parents were gone. She wanted Holly to share them with her, promising that no one would hear about them until after Mom's passing. Holly simply told Gloria that she had decided she wasn't going to pass them on to anyone. She felt that it was appropriate to let these stories die along with her, and nothing could convince her otherwise. The stories would be lost, and there was nothing we could do to convince Holly to let us record them.

I was not at all happy to know that Holly would be taking these remarkable stories to the grave. Fortunately, not all of them were lost; a small handful was shared over the years with the older siblings who had also lived through those years. They usually came out when we'd gather for reunions, weddings, and other family events. Some were shared by Holly, while others were shared by our aunts and cousins. Of course, they were never shared in the presence of Mom or Dad.

Over time, this small collection of stories was repeated at every gathering. As I grew older, I purposely declined invitations to go and play with my cousins. Instead, I'd park myself out of sight but within hearing distance so I could hear some of what my older siblings and aunts were talking about. Later, recalling some of the stories, I realized many of them provided me with enough clues to begin some in-depth research on my own. Without Holly or anyone else knowing what I was doing, I soon discovered some key facts and details.

Not all was lost with the passing of my oldest sister. With the advent of affordable DNA testing, I hoped we might be able to find out why Holly felt so strongly that these stories needed to be taken to the grave.

Fortunately, some stories never really die. When suppressed, they simply find other ways to manifest themselves.

2

THE STORIES

When Holly died, we lost a big part of our family's early history—the part that took place before the last three kids were born. She was the oldest child and, as such, witnessed the greatest portion of our family's early history. Also, as the oldest, she took charge when, for whatever reason, Mom and Dad were not able or willing to do so.

Despite the loss of Holly's knowledge, some events were shared by other people many years before her death. My sisters Judy and Gloria were born only a few years after Holly, so they lived through most of those early years, as well. Over the years, Holly had shared some of the stories she remembered with Judy and Gloria, so some of the stories survived through them. In fact, not only did they survive, but they were told repeatedly, not only to us younger kids but to the nieces and nephews. In this way, without realizing what they were doing, they made sure that a handful of our stories would survive the loss of several family members. These are the ones that helped open the door just wide enough for me to see some of the events that helped to shape our family's history. They also allowed me to realize that something wasn't quite right. I had begun to wonder what really went on during those years—and who were these people whose names kept popping up?

Following are a few of the stories that played a key role in the events that would come to a head and begin the process of discovering the real nature of our family, a story we would find almost too incredible to believe.

THE ACCIDENT

One important event took place when I was about fourteen years old. Four decades later, this very event would set a quest in motion to discover answers that would entirely reinvent our family's history.

This event was important not just because it was the first time we would come face-to-face with the possible death of an immediate family member, but because of a secret it revealed, and what else it would set in motion. I have no doubt that much of what we discovered during those days involved some of the stories Holly was so unwilling to talk about.

It all started when my brother Neil was sixteen years old and left home to spend the summer working on a friend's farm in Indiana. Things had gotten rough for Neil, living at home. He was not getting along with my parents. Neil was a typical teenager, and my parents weren't handling it well. He was having an especially hard time with Mom; those two butted heads often, rarely getting along with each other. Because of that, Neil and my parents thought it might be good for him to spend a summer away from everything that had been happening at home. It would also be a good way for Neil to learn how to work hard and take on some adult responsibilities. My dad had grown up on a farm, so he knew what Neil was in for. He believed that working on a farm was just what a young man needed to grow up and find himself; plus, all agreed it would be an added benefit for everyone if Neil had some time away from Mom.

One evening, about a month after Neil had left home, Mom and Dad got the kind of call that every parent dreads. They were told that Neil had been involved in a terrible accident. He had been driving a tractor on some steep terrain when it tipped over, crushing him beneath it. He was rushed to the hospital in hopes they could save his life, but things didn't look good. My parents were told to drive the two and a half hours to the hospital to see Neil for what could be the last time, to say their good-byes before he died. It was a long and painful drive that gave my parents time to ponder the possible death of their son.

When Mom and Dad arrived at the hospital, they found that the hospital had sent out a call throughout the small farming community, asking for blood donors. Of course, Dad was right at the head of the line, prepared to donate all that they needed. A few hours later, Dad was told that they couldn't give his blood to Neil because his blood was not compatible; they knew that his body would reject it. Hospital staff said they were willing to trade out Dad's blood for other blood they had available, and they would give Dad credit for the donation. The fact that Neil's blood was not compatible with Dad's blood was not conclusive evidence that he was not Mark's son, but Mom didn't know that. I learned later that when Mom thought the secret was out, she admitted to Dad that Neil was not his biological son.

With a single blood donation, the dynamics of our family had changed forever. That night, in a small town far away from home, faced with the very real prospect of losing one of his sons, Dad learned that Neil was not his biological son.

After Dad's death, I spoke with one of his war buddies. He told me that even before Dad found out who Neil's real father was, he had suspected that something might not be right. Neil looked so much like his biological father—who was a part of our lives—that Dad had suspected it was more than just a coincidence. This man showed so much interest in Neil that it wasn't entirely surprising Dad would be suspicious. Now, he knew for sure. Whether Neil had lived or died that night, it appeared that Dad had already lost a son.

It is fascinating to think about what stories are bound up within our blood. For as long as people have lived on this Earth, they have had painful and unsavory secrets to hide. It used to be that there was little or no evidence of illicit actions that took place, but in recent years, science has been able to reveal the truth. The evidence we try to hide is recorded within the very blood that keeps us alive. This was only the first of several times that blood betrayed one of our family members, revealing secrets that some tried to hide, and others never knew existed. We would learn repeatedly that with today's technology, family secrets are not as easily hidden as they used to be.

I don't remember anyone talking about what happened after Mom and Dad came home from visiting Neil in the hospital. It was simply understood that no one would bring the matter up for discussion. I was young enough that the details weren't of much interest to me, but I have no doubt that Mom and Dad discussed it at great length in private. It was shortly after this event that Dad moved to an upstairs bedroom, while Mom kept the bedroom downstairs. I'm sure this was not a coincidence.

I didn't think anything of this new discovery. I was more concerned that Neil was hurt so badly that he could possibly die. The thought scared me. We had never lost a close family member, especially someone as close as a brother. I couldn't imagine what our family would be like without Neil. It was that simple.

When I was told what it really meant—that Neil was not Mark Anderson's son—I didn't fully appreciate the consequences of that announcement. I just thought that maybe Neil had been adopted, and that Mom and Dad didn't want us younger kids to know about it. I did think it was kind of cool that my older brother might have been adopted. Thinking back to that time, I don't understand why I didn't ask more questions about it.

Neil's biological father was a family friend named Dennis, who lived with his wife just a couple blocks from our house. I had been to Dennis's house with Mom and Dad many times. Dennis was one of Dad's card-playing buddies, a fellow World War II vet. I think Dennis knew all along that Neil was his son. When Neil was only about thirteen years old, Dennis hired him to work with him during the summer months, delivering baked goods throughout the Chicago area. I always thought this would be the perfect job for a kid like me. I was so jealous of Neil; not only did he get paid a decent wage, but he also got all the goodies he ever wanted. Dennis was more than happy to indulge him.

The two of them spent the summers together, traveling the roads in the delivery truck, talking and enjoying each other's company. Dennis bought Neil a brand-new bicycle for his sixteenth birthday. It wasn't some cheap bike, either; it was a high-flying, top-of-the-line Schwinn, bought from the show window of the local Coast to Coast hardware store. This man never did anything special for me or for Tim. I must say that I was more than a bit jealous of all the attention and kindness Neil received from Dennis. I couldn't figure out why he was being treated so special. Now it all made sense. He knew Neil was his son, and he was trying his best to be a father figure without letting Neil know that he was his real father. Looking back, I think it's great that Dennis could spend so much time with Neil. I know Neil enjoyed the attention, and I'm sure Dennis enjoyed getting to know his son, and watching him grow up.

Once I discovered this secret, something else made sense. One Easter, when Neil was twelve and I was ten, Mom made it a point to buy us both matching suits. She cut our hair in the same style and had us pose side by side in the same stance for Easter pictures. Then she took the photos and showed everyone she knew, commenting on how much Neil and I looked like twins. With only fifteen months between us, we were close in size. Personally, I never understood why she tried so hard to make it seem that we looked alike; in fact, no one could understand it. We didn't look anything alike. No one in their right mind would have thought we looked remotely like brothers, let alone twins. Where was she coming from with such a silly claim? Given what we had learned after Neil's tractor accident, I'm sure her intention was to try and convince people that we were brothers in the fullest sense of the word.

I am not sure when Neil finally found out that Dennis was his dad. Every family has their own channel of information distribution. In our case, news of Neil's biological father probably went from Mom to some of her sisters. From there, I'm sure our aunts shared it with their husbands. One or more of their kids probably overheard everything. They, in turn, told Neil that they knew a secret about him.

I eventually asked Neil when he had found out about Dennis, and he simply said that he didn't remember. To this day, Neil has no interest in talking about Mom, or what she and Dennis did. Mom's relationship with Dennis isn't the only reason Neil harbors hard feelings toward our mother, but I don't doubt that discovering this secret played a huge part in what took place in the years following the accident. I know how important it is to the person who has been offended to let go of the offense, but some events take a lifetime to let go of. I don't believe that Neil has reached the point where he can forgive Mom yet. Knowing Neil, it may not happen in this lifetime, but for his own peace of mind, I hope he can let go of it soon and not carry that pain with him into the next life.

More than forty years after this event took place, I was talking with Mom. She told me that Neil had never forgiven her for what she did with Dennis. She couldn't understand why he was holding on to this. I could only tell her that time would take care of these things. I'm not sure it was any comfort to her, but to be honest, I don't think there was anything I could have said that might have given her any degree of comfort. I know she carried this pain with her to the day she died.

I'M SO MAD AT DAD

A second puzzle piece fell into place when I was about thirty-five years old. It had to do with my sister Holly. As I've mentioned, Holly was the oldest of our family's nine children, and was always considered the matriarch of the family. Mom seldom embraced the responsibilities of that role with much zeal. At a time when women typically stayed home and raised their families, Mom always had a job that allowed her to get out of the house and have some extra money. Since Dad was a traveling salesman, he was gone a lot, too. This was a time in Dad's life when he was drinking heavily, so it wasn't like he stepped up and took over the responsibilities of raising us.

So, in the absence of a healthy, nurturing mom or dad, the three older girls stepped in and made sure the family was cared for and didn't disintegrate into total chaos. Mom never completely gave up her responsibility of raising her children, nor did the oldest three girls ever take over complete control of our care. I have many memories of Mom doing the things moms do, but I also remember many times when she would run off with Dad for a night out, or when she would go and do something with her sisters. Then, she would leave us in the care of the oldest three girls. Occasionally, Mom would be gone for two or three days at a time, leaving my older sisters in charge.

There were many times when Holly's actions ran into direct conflict with what Dad thought she should be doing. Holly was a very strong-willed teenager. She was not one who could be easily pushed around, even by our parents. I remember many times when Holly and Dad got into some serious fights. Dad never appreciated it when Holly tried to boss him around or go against his will. He wasn't about to have his daughter usurp his role as head of the household, even though he was away from home most of the time. Family roles didn't mean much to Holly, probably because she felt that Mom and Dad so obviously neglected those roles themselves.

One day, after Holly had had an especially rough run-in with Dad, she came into Mom's bedroom and voiced her frustrations. "Ohhhh, I hate Dad," she told Mom. "I wish he wasn't my father!" According to Holly, this was the moment Mom chose to tell Holly that Mark Anderson wasn't her biological dad.

I have no clue why Mom chose this time to tell her. Holly never shared any more details than that about the conversation. I don't know if Mom told Holly who her biological father was at this time, or if she provided those details years later, but Holly eventually discovered who her biological father was: a man we all knew very well, named Sam, who, we would soon discover, had played a much bigger role in our family than any of us realized.

Learning that Mark was not her real father only made matters worse between Holly and Dad. She had never been close to him in the first place, but afterward, she was even more distant. Finding out that she was not Mark's daughter removed any real obligation on her part to try and develop a relationship between the two of them. She didn't share any of Mark's genes, so it seemed she never felt a need to make things work between them. There was no longer any real sense of being bound by familial duty, or blood.

Holly told me that she was Sam's daughter when I was about forty-five years old. She and I were talking about our family history, and she thought I might be interested. It didn't really come as a surprise to me. I remember Mom sharing a story with me that helped me see how this could have come about. Before Mom married, she was dating two men: one was Holly's biological father, Sam, and the second man was our dad, Mark Anderson. She loved them both and was seeing both men. Mom eventually became engaged to Sam, and the two of them planned on getting married not long after Sam proposed.

Mom said that she and Sam were at a dance in a neighboring town when she told him she wanted to dance with Mark. Sam made it very clear that he was against this and demanded that Mom never dance with Mark again. They were engaged to be married, and it would not look proper for her to be out on the dance floor with her old boyfriend. It would be an embarrassment to him and his family, and he demanded she never even consider doing such a stupid thing again.

My mom was one of those people you could never say no to; nor would she tolerate anyone telling her that she was doing something stupid. To do so was to put your own well-being in jeopardy. She never took orders from a man, not even her own fiancé. If you asked her to do something, she was fine with that . . . most of the time. But if you ordered her to do something, there would be hell to pay. She never allowed a man to tell her what she could and could not do, plain and simple.

Mom and Sam got into a big fight right there on the dance floor. Mom told Sam that he was never to order her to do anything. Of course, Sam would have none of that. He was raised with the old-school idea that a woman was a man's property, and that she was obligated to follow her husband's orders with unquestioning obedience. Mom was incensed. She told Sam that the engagement was off and that he could go to hell. Not only that, but she was going to marry Mark Anderson. She would never give Sam the satisfaction of claiming her as his "property."

Soon afterward, Mom married Mark; however, with time, Mom's feelings changed toward Sam. She eventually got over her anger and wanted to keep the relationship alive. I wonder if she regretted marrying Mark, or if she was simply still in love with Sam. Whatever her feelings toward Sam, she knew she could never live under his old-fashioned belief that women should be subject to their husbands' rule.

Shortly after Mom and Dad were married, he enlisted in the army. He was shipped off to serve in Texas, eventually making his way to Europe. Throughout his four years in the military, several months could pass between his visits home, which meant that Mom was on her own for much of those years. It was during this time that Holly was conceived.

THE SISTERS TALK

A third puzzle piece came to light several decades after my sister Carlee was born. Carlee was the ninth and last child in our family. She was fifty-three when she found out she was not Mark Anderson's daughter. She overheard the three oldest sisters talking about her, saying that her real father was not Mark Anderson, but rather some young man who had worked for Grandpa at the family business. When Carlee asked the sisters about it, wanting to know if it was true, they simply told her to ask Mom. They were sure that Carlee wouldn't try to approach Mom about something this sensitive, but as it would turn out, they were wrong.

It appears there was a handsome young man from Germany who was working at the family shop back then. The older girls really liked him because, according to them, he was handsome, a lot of fun to be around, and he had a car. In our small suburb we didn't have a lot of people moving in, and it was a rare thing indeed to have a young, single man come to town—one who was good-looking and foreign, no less. His name was Gerhardt, although no one called him that; he preferred to go by "Peter."

My older sisters were quite taken with Peter's good looks. He knew my father very well because he worked with him in the same division of the family company. He would often come over to our house and talk with my older sisters, one of whom enjoyed flirting with Peter, which suited him just fine. He loved all the attention that came his way. Peter would take some of the girls for rides in his car to the east side of town, where there was a lake perfect for swimming.

After hearing these stories, including the possibility that Peter was her father, Carlee finally came right out and asked Mom if it was true. What amazed everyone is that Mom confirmed it without batting an eye. She didn't ask how Carlee knew about Mark not being her father; she just told her all the facts and left it at that. No apology, no explanation—just what Carlee believed to be the truth.

Of course, when the sisters found out that Carlee had asked Mom about this, they were mortified. It was one thing to talk to Carlee about this guy and share their stories with her, but for Mom to know that they all knew about Peter—and that they had even gone so far as to tell Carlee about him—was crossing the line. They waited for Mom to come after them for telling Carlee this secret, but it never happened. Mom never said a word about it to the older girls.

This surprised us all when we heard about it years later, as this was not how Mom typically operated. We never did learn why Mom didn't say anything to anyone once she knew the secret was out. Maybe she was simply relieved to know the truth had finally been revealed and she didn't have to hide it anymore. It's hard to say why things turned out the way they did, especially with Mom.

When Carlee confronted Mom about Peter, she didn't seem at all upset that Carlee knew the secret. In fact, she told her the whole story, explaining that it had happened after she had been out drinking with Dad on a Friday night. Dad was playing cards and Mom knew he would be at it until late into the night. After waiting as long as she could, and having a few too many drinks herself, she decided to ask Peter for a ride home. Peter was happy to oblige. No one knows any of the details about how things got started from there, but apparently, this was not a direct trip home. It included an unplanned diversion and came with some unexpected surprises.

Apparently, Peter took advantage of his good looks on more than one occasion, because we found out a few years later that Carlee was not the only memento of Peter's stay in town. We discovered that Peter had also been dating my aunt Ruth, Mom's sister. Nine months after that date, our cousin David was born. (So not only did Carlee find out that Mark Anderson was not her real father, but she also found out our cousin David was also her half-brother.)

It appears that most of the adults in the family knew what Peter was up to with Grandpa's daughters, because he ended up firing the young man. A few days later, Peter was nowhere to be found. Apparently he left town before Grandpa's friends could take care of the matter in their own way. It was bad enough that he had gotten one of Grandpa's daughters pregnant, but two of them? This guy was a real liability to the family's reputation.

By the time Carlee learned this information, Peter was long gone. She had no idea who he was, where in Germany he came from, or why he'd been in the Chicago area in the first place. No one remembered any firm details, so to this day he remains a mystery.

With each new discovery, we were finding out that discovering one's family history is far from boring.

3

CREATING THE STANDARD

For years, Tim and I had considered the idea of creating a "genetic standard" by which each of us nine children could compare our DNA against the DNA of our parents. This would be the only way we could know for sure who was and who was not biologically related to our parents. By 2007, I knew that the only way we could really prove or disprove the facts of what we were learning and make certain the information was accurate was to make use of DNA testing services.

Ancestry.com, the world's largest online family history resource, was leading the field in providing the public with affordable AncestryDNA testing services. In addition, they had the biggest database of genealogies linked to their DNA test results in the industry. A friend who worked for AncestryDNA.com explained that AncestryDNA had just recently started to make DNA testing available to the public at a very reasonable price. Their original purpose was to make it possible for people to see a breakdown of where their ancestors came from. As they refined this test and built up their database of users, they were also able to provide consumers with a good idea if a person listed on their online family tree was related to another person in a different tree.

Two other companies, 23andMe and MyHeritage, were also developing their DNA testing products and creating their own DNA-linked genealogy databases to grab their share of the consumer market, as well. Genealogists around the world were seeing that DNA tools were taking genealogy and family history to a whole new level of sophistication. I've witnessed this personally after more than forty years as a genealogist, oral historian, and family history blogger, working twenty-six of those years for FamilySearch International. Tim and I knew that with the help of these companies, and a few other options open to us, we were ready to consider taking the next step toward finding some of the answers to our questions. We were ready to begin moving toward creating a reliable standard against which any of our siblings could be tested; all we had to do was gather a sample of DNA from both of our parents.

Various media outlets were just beginning to air stories about people who had taken DNA tests only to discover they were not the children of the fathers (or parents) who had raised them. The public was fascinated. Over time, more and more of these stories started to pop up on the Internet, television, and several other media outlets. Books on this topic were not common in 2010 when Tim and I gathered our first DNA samples, but by 2016, several had been published. It was becoming more common to hear of someone who had tested their DNA with one of the popular testing companies only to discover a surprise that changed their life.

Richard Resnick, CEO of Genome Quest, presented an interesting TED Talk in September of 2011, when the genomic revolution was still in its relative infancy. He stated that "in any given audience, an average of 1 percent to 3 percent of the men in that group are not actually the father of their child."[1] That comment got a definite response from his audience. Resnick's TED Talk was all about how genetic testing was changing our world, faster and more dramatically than any of us dreamed possible. For many, it would be the discovery of a new parent–child relationship.

Tim and I both knew that DNA testing was the only way we were going to find the answers we were looking for—the only way to sift truth from the misinformation that was being passed from one family member to another. After Neil's accident, we felt very confident that at least one out of the nine of us was not Mark Anderson's child. In addition to Neil, there was also a high probability that Holly and Carlee were not Mark's daughters, because Mom had admitted as much. However, we knew that Mom was not always honest with us. In fact, when it came to matters of family history, she probably made things up more often than she told the truth.

Of course, not everything she did was in an effort to deceive. You have to remember that the actual events took place many years—in some cases, decades—before any of us started asking questions. It is easy to forget things or mix up the facts; details often evolve and change with time, sometimes intentionally. And there was a very real possibility that some of my siblings were simply telling us what they had heard from others, who were sharing information based on rumors and suppositions. So even though it had not yet been 100 percent confirmed that Holly and Carlee were not our full sisters, both of them felt fairly confident we could prove this, one way or another, by testing their DNA. With one-third of the family most likely not biologically related to Mark Anderson, I had to wonder if there weren't more surprises associated with my other siblings hidden away in our family's story.

The cost of DNA testing had previously been a barrier for us when it came to moving forward with our plan. But now, we knew it was time to do something.

Before we could come to any positive conclusions about who belonged to whom, we had to resolve a few key issues. Probably the most important was deciding how we were going to get DNA samples from our parents. In families where no one is hiding any incriminating secrets, this probably wouldn't be a big problem. You would simply get a cheek swab from each parent and you'd be ready to go. But in our case, we were pretty sure that Mom and Dad both had some very big secrets to hide. How could we get DNA samples without raising suspicions that we were asking serious and intrusive questions about whether or not Mark was really our biological father? We couldn't come right out and ask them for samples. We also had to look at this from a moral point of view. Would it be ethical to try and get samples without telling our parents what we were planning to do with them?

Judith Russell is a world-renowned genealogist and legal expert who has spoken at genealogical and family history conferences around the world. In an article she posted on her blog on November 12, 2012 (about the time we were trying to gather DNA samples from our parents), she provides some excellent information about the legality and ethics of acquiring DNA samples without consent:

> By 2012, 11 states had some form of comprehensive statute requiring consent for genetic testing, 15 states and the District of Columbia had no laws on DNA collection or disclosure at all, and most of the remaining states only limited DNA testing in certain contexts, such as insurance or employment.
>
> So, for the most part, you're not committing a crime if you manage to snag a DNA sample without somebody's consent. Getting it tested by a genetic genealogy company, on the other hand . . . well, you'd certainly be committing a fraud if you tried. All four places where I've currently tested—23andMe, AncestryDNA, Family Tree DNA, and National Geographic's Geno 2.0 project—require that the sample be submitted by the person whose sample it is, or by someone with the legal authority to consent for that person (such as a parent or guardian on behalf of a child). For example:
>
> - 23andMe's terms of service require that "You are guaranteeing that any sample you provide is your saliva; if you are agreeing to these TOS on behalf of a person for whom you have legal authorization, you are confirming that the sample provided will be the sample of that person."
> - At AncestryDNA, "You represent that any sample you provide is either your DNA or the DNA of a person for whom you are a legal guardian or have obtained legal authorization to provide their DNA to AncestryDNA."
> - Family Tree DNA explains, repeatedly, in a number of contexts, that "Even if you paid for the test of a friend or relative, they need to be the one to consent . . . we ask that you practice ethical testing and kit conservatorship."

> So . . . why should we care? What's the harm, anyway? After all, that cousin isn't doing anything with that DNA sample, and I'm not going to misuse the information, right? The hitch is that the ethical line is really clear: yeah, we want the information, but it belongs to the other person who has the right to say yes or no. It's the old Golden Rule again: if we want the right to say no to an intrusion in our lives, we have to acknowledge that others—even our doggoned cousins with the DNA we really, really need to break down that brick wall—have the right to say no as well.[2]

Given this information, depending on the state one lives in, it may or may not be legal to acquire a DNA sample without consent. Regardless, it would not be considered ethical to obtain a DNA sample from a family member without informing them of what we were doing. Yet, in this case, finding out if we were Mark's biological children had very big medical implications for us and for our future posterity, not to mention emotional implications, as well. Gastrointestinal cancer and diabetes were both high-profile diseases on Mark's side of the family. By this point, two of our siblings had already died, both from cancer.

We had to ask ourselves: What was more important—ethically crossing the line by obtaining DNA samples without telling our parents, or our right to know the truth about the medical history associated with our biological bloodlines? The decision was not going to be an easy one, but we had to make it, and soon. Both of our parents were in their eighties, and we knew they wouldn't be with us much longer. Still, we decided to hold off for a while and see how things went. Maybe something would come up that would help us in our quest to get DNA samples.

When Tim and I first seriously considered how we would start gathering DNA samples, we weren't sure we wanted our brothers and sisters to know what we were doing—at least, not right away. We knew that our oldest sister, Holly, if she had been alive, would have put an immediate stop to our efforts. She had known a lot more than any of us, so we didn't know how far we could, or should, go with this.

At this point, I told Tim that we needed to get started with our efforts to create a baseline standard against which we could all be tested. Once we had this, we could wait until Mom and Dad were both gone before any of us started any serious efforts to be tested. This way they would be spared any possible humiliation from knowing that their children suspected there were secrets in the family. We had no interest in doing anything to embarrass them.

The second challenge was deciding which company we were going to use to test our DNA samples. AncestryDNA, 23andMe, and MyHeritage were not yet offering DNA tests to the general public on a large scale. Our goal at the time was simply to prove whether or not we were the biological children of Mark Anderson. I already knew my ethnic background, so companies who said they would soon be offering DNA tests to establish ethnicity didn't interest me. Most of the "at-home consumer DNA testing companies" were still building up their databases to make their products more marketable, so I wasn't confident it would do much to connect me to other relatives. Besides, I knew all my cousins. Given our goal of confirming that Mark was our biological father, connecting with third or fourth cousins just didn't interest me.

Even though we had no reliable proof of any kind, Tim and I had no doubts that we were Mark's biological children. We look like our dad in so many ways, and we both have many of the same mannerisms. I also felt reasonably sure that some of our other sisters were biologically tied to Mark Anderson. But I felt strongly that we needed to allow anyone who had doubts to have a way to get the answers they needed. For that reason, we decided to bypass the three popular consumer DNA testing companies and go with a private lab.

I knew that going with a private DNA testing lab would be more expensive, but given what we were looking for, we couldn't afford any mistakes. Plus, I wanted to have our family's DNA information on file so we could run additional tests—many of which are not offered by the other consumer testing companies—if we ever decided we wanted, or needed, to learn more. Tim and I both felt like we needed to have a genetic testing company that had experience, a good track record of success, and was willing to talk us through the process, helping us understand the results and implications of what we were learning with each test.

Once again, life got in the way, and we laid our plans aside. We decided to tackle it at a later date. Things didn't progress past this planning stage for a couple more years. Then one day I got a call telling me that our dad had passed away. He was eighty-eight years old and struggling with advanced diabetes when he died, so it didn't come as a big surprise. But as is so often the case, even when it's expected, hearing that a loved one has died still comes as a shock.

Dad had mentioned to me recently that he was ready to go—that he wasn't worried about dying. During the last few years of his life, I had called him every Sunday night to talk with him. In his old age, he had become more open to things; I was hoping he might say something about what we had learned up to this point, or indicate that it would be okay to approach him. Every time I called, I would pepper Dad with a few questions about our family's history. I'd ask him about dates that he was away from home when he was in the army, and what he remembered about the birth of each of his children. I was always on the lookout for something I could use to broach the subject of our births, but nothing ever came up. He never gave me even the slightest hope that he would be okay with me asking him these kinds of very personal questions I needed answers to.

After about a year of this, I began to wonder if Dad had the slightest inkling that any of his kids (except for Neil, of course) might not have been fathered by him. It had never dawned on me that Dad might not have known that any of his children were not his. This insight added a new dimension to our investigation. How much did Dad really know about all of this? Did he have any clue that Mom had been involved with so many of his close friends?

There were times when I asked myself whether I was being entirely inappropriate by pursuing the issue of our paternity. After all, kids don't do this kind of stuff to their parents, do they? Was I being completely and utterly disrespectful toward my parents by butting into such incredibly private matters? But each time I wondered about these things, the thought would come to me that these matters were extremely important to each of us children, and to future generations. As genetic research opens new windows into nearly every aspect of our lives, we honestly need to know which family tree we are a part of when it comes to pursuing medical information. Barking up the wrong tree could have life-threatening consequences.

These were very difficult times for me. Although I had mixed feelings about all of this, something kept pushing me on, letting me know that we had to learn the truth about our identities, and what role each of us played in this family of ours. I must admit that I had many sleepless nights as I struggled to come to terms with what needed to be done.

We were in the funeral home for Dad's viewing when Tim pulled me aside to remind me that if we were going to get a DNA sample from Dad, this would be our last chance. It was now or never. In four more hours, he would be buried six feet under, as good as a million miles out of our reach. My first thought was, How in the world are we going to get a cheek swab from a dead man? After all, Dad had already been embalmed. I wasn't familiar with how they prepare people for burial, but I was sure they did something to ensure that the person's mouth didn't suddenly open during the viewing—an adhesive or wire device of some kind. I didn't know if that was true or not, but if that was the case, getting a cheek swab was pretty much out of the question.

Tim reminded me that if we couldn't get a cheek swab, some hair samples with the hair follicle still attached would do. He thought about the problem for a minute and then left the room. I didn't know it until after the viewing, but apparently Tim spoke with the funeral director and explained our situation. He didn't go into detail, but asked if the mortician would be willing to pull out some of Dad's hair for us, after the viewing was over. I think he told the director that we were gathering DNA information about all of our family members so we could compile a family medical history, or something like that.

I'm sure the poor funeral director must have thought we were very disturbed, asking him to pull a handful of hair from our father's corpse. As I think about it now, it does sound very creepy. But as Tim and I were leaving the funeral home, the funeral director passed Tim an envelope. Tim quietly tucked it into his suit-coat pocket. When I asked him about it later, Tim confirmed that it was Dad's hair.

Now that we had some DNA from Dad, we needed to get some from Mom, too. She was about eighty-five years old at the time and in good health. We knew she could be around for several more years. Waiting until her funeral was simply not a good option.

Several months after Dad's funeral, I got a phone call from Tim. He told me that he had a top-quality sample of Mom's DNA. I was surprised to hear it, and asked how he'd gotten the sample. He said that after trying to figure out a way to get a good cheek swab from Mom, he decided he would just ask for one. Tim used basically the same tactic he'd used with the funeral director, telling Mom we were gathering DNA samples from family members to compile a health history, which was true. We wanted to include her DNA when creating the genetic standard for our family so that in the future, we could use that standard to test for medical issues, in addition to answering our paternity questions. Mom never batted an eye; she was happy to help. It was just that simple! I had to laugh. I can always count on Tim to do what it takes to get something done. The guy is a genius!

We put the samples in the freezer, and there they stayed for nearly two years. We kept putting off the testing of our own DNA samples, thinking we'd get it done when we had the money to pay for the tests. This wasn't a pressing issue for Tim or me; we were both confident that we were Mark Anderson's sons. We were doing this so we could provide a comparison base against which our brothers and sisters could be tested, if they so chose.

Now that we had the DNA we needed, we still needed to find a place to get the testing done. Our number one question was which of our brothers and sisters were fathered by Mark Anderson, the man who raised us. Our mother had admitted that Mark was not Neil's father. (We still needed to do DNA testing on Neil to be 100 percent sure, but that would have to come at some future date, if Neil would agree to being tested.) None of us were 100 percent certain that Mark was Holly and Carlee's father, but we thought the chances were very high that he was not. Our second question was, if Mark Anderson was not the father, then who was?

Actually, at that time, we didn't expect that we would be the ones to find these answers; we would simply provide the standard and then let each sibling choose whether they wanted to pay for testing to find out for sure. Those who discovered that Mark was not their father would have to decide if and how they would take the next step to find out who their biological fathers were, and how much they would be willing to pay to learn that information. As far as I was concerned, that was a bridge we'd cross once we got there. For now, Tim and I needed to focus on creating the standard and then letting our brothers and sisters decide if they wanted to go any further.

We didn't want a cheap company with little experience or with a shoddy reputation to do this job. But with two parents and nine children, we couldn't afford to pay several hundred dollars for each person to be tested. We wanted to make sure that whoever we hired was AABB (American Association of Blood Banks) and CAP (College of American Pathologists) certified. These certifications generally ensure that the companies will provide the best possible results. We wanted to know the truth, but we didn't have a lot of money to work with, nor did we have a lot of Dad's DNA material to spare. We also weren't sure that we could get more DNA from our mom, so we needed to be very careful about who we selected to do our testing.

Once again, the answer came from Tim. His good friend Jack Anderson (no relation) owned a genetic testing firm called Andergene Labs in Oceanside, California. The lab had a very good reputation for its work with paternity and other genetic testing services for people around the world. Tim had talked with Jack about what we hoped to do with our parents' DNA, explaining that we wanted to learn whether our siblings were biologically linked to Mark Anderson. Jack was intrigued by our case and gave Tim some advice about how to move forward. He said that if we were interested in testing the samples we had, he would make it affordable for us. This was the man who could help us create the standard we were hoping to establish.

I told Tim that we had to move ahead on this and take Jack up on his offer. I decided that I would pay to have Mom and Dad's DNA samples tested, and while I was at it, I would have mine tested as well. Things were starting to fall into place; I was excited to finally have our standard in place. I wasn't concerned about my own DNA showing anything unusual. I knew I was Dad's son. Just look at any picture of me with my dad and my uncles, and you will see that there is no doubt I am an Anderson all the way. Nor was there any doubt that Tim and I were full brothers. For years, if anyone called our home, they could never tell whether they were speaking with me or with Tim. I can't even begin to count the times that people told us we sounded exactly alike. Not only did we sound alike, but we looked alike; Tim looked even more like Dad than I did.

So for us, the issue of finding out whether we were related to Mark Anderson wasn't even up for consideration. However, I did want to have my DNA tested to see if there was anything interesting that Jack could tell me. Although this wasn't one of our main research goals, I thought it would be informative. Besides, it would be good to have my genetic information readily available if I ever needed to determine if there were any diseases I might be susceptible to, or anything serious that I should be aware of. I thought it would be a good investment.

Tim gave Jack the samples he had gathered from our parents, and I provided Jack with some of my own DNA. I was surprised by how simple and painless it was to provide my material for testing. All he needed was a couple of cheek swabs. This was a matter of taking two long cotton swabs and rubbing them along the inside of my mouth. Each cheek was rubbed with a different cotton swab; it was that simple. No blood was needed, and there was nothing that had to be poked or inserted into my body. With the cheek swabs ready, we were good to go. The adventure was moving to the next level. It was getting exciting now!

Jack told us that it would be a month or two before he could provide us with the results of our tests. I wasn't concerned; after all, these samples had been sitting in Tim's freezer for nearly two years. I could wait a couple more months.

We now had a way to get the answers to our first question, about who is, or is not, biologically related to our dad, Mark Anderson. Jack was going to provide this information for us. The answers to our second question—if any of our siblings discovered that Mark Anderson was not their biological father, how would we find out the identity of their fathers—would have to come from online genealogy database companies like AncestryDNA, 23andMe, and others. For now, we were focused on answering our first question.

I was at work when I got a call from Jack. He had the results of our DNA tests and was ready to explain what they revealed. I was excited to hear what he had learned about Mom and Dad, and to see whether he had found any unusual health issues from my own DNA sample. While the main purpose of this test was to establish the standard against which we all could be tested, regarding our biological link to our parents, my interest was to see if I had inherited any physical traits or genetic tendencies that I needed to be concerned about. With this test, the standard would be in place and I could leave the rest up to each individual sibling to decide how far they would take this journey. It would be the scientific, unquestionable proof of parentage. For me, I was done. My story was set in stone, assuming Jack did not find any genetic markers for worrisome health problems.

Jack assured me that the samples Tim had stored in the freezer were still viable, and that he had everything we needed to create a good baseline against which to compare other DNA samples from our siblings. I asked Jack if he had found anything interesting in my DNA sample. Since Dad had struggled with diabetes starting in his early forties, and many of my Anderson relatives had died of both diabetes and stomach cancer, I was interested to learn whether I had inherited any of those traits that would make me a high-risk candidate for either of these two medical conditions.

He told me that yes, indeed, he had found something.

I knew it; it was time. I was ready to hear the bad news about which of those two diseases I was genetically prone to—but I was not ready to hear what Jack was about to tell me.

The good news was that I didn't have any markers for early-onset diabetes or any unusually high markers for stomach cancer.

The bad news was, Mark Anderson was *not* my biological father.

4

SURPRISE! DEALING WITH THE RESULTS

I remember when I was about fifteen years old, I got into a fistfight with a kid in school. I don't remember the details of how the fight started, or why I would have been stupid enough to allow myself to be sucked into this fight, but I did. The other kid delivered a gut punch that took my breath away. It felt like the longest twenty seconds of my life. I was panicking because I couldn't get my lungs to suck in air. I was sure I was going to die.

When Jack told me that, according to the DNA sample I had provided him, Mark Anderson was not my biological father, I had that same sensation of panic. I couldn't catch my breath. There was literally twenty seconds or so of silence before I could get my lungs to take in enough air to respond. At first, I thought I'd simply misunderstood what Jack was telling me. I asked him to repeat what he'd said, and to do it slowly, so I could take it all in. I had a small degree of hearing loss, a genetic trait common on my mother's side of the family, so I thought that perhaps I hadn't heard him quite right.

He humored me and slowly repeated what I had heard him say the first time. Jack sensed my shock, and was trying to be as reassuring as possible. I have no doubt that he'd dealt with this kind of response countless times as he delivered the results of paternity tests. I asked him to repeat the results a third time, just to make sure. He told me that these tests are never 100 percent accurate, which gave me a temporary sense of hope. However, he went on to say that he could assure me that the tests he runs are 99 percent accurate, with 1 degree of variability. So while he was 100 percent sure his test results were accurate, because statistics work the way they do, he had to legally state there was some degree of uncertainty that his testing might not be able to account for. That's standard practice with this business. They say it's 99 percent accurate only because in theory, there is no perfect test. As far as we were both concerned, the results of this test were 100 percent reliable. No mistakes were made. My dad was not my dad—at least, not biologically.

On September 7, 2018, Brianne Kirkpatrick, a genetic counselor, DNA consultant, and founder of the Internet site Watershed DNA, appeared on NBC's *Megyn Kelly Today* show. She stated that by 2017, more than 50 million DNA kits had been ordered from online DNA companies. She estimated that 5 to 10 percent of those tests had revealed cases of "non-family" relationships—in other words, where one (or both) parents were not the biological parent of the person being tested.[1]

That's an amazing statistic! This means that out of a random group of one hundred people, anywhere from five to ten people within that group could have a parent who is not biologically related to them. That's shocking! Assuming that this statistic is true, then in 2012, when I got that first phone call from Jack, approximately 15.5 million to 31 million people in the United States faced the same situation I was facing. Today, DNA testing is rocking the world of many of those people by exposing the truth of their paternity. It's been said that misery loves company. That may be true, but I was not comforted by knowing that more than 15 million people shared my story.

When I started this effort to find answers to our family's questions of paternity, I had no doubt that I was Mark Anderson's son. The possibility never even entered my mind that I might not be related to him. Both Tim and I were sure we were Mark's sons. Now I'd found out that I was fathered by another man, maybe even a complete stranger. If true, this meant that Tim and I were most likely half-brothers. What just happened? With one phone call, my world had turned upside down. I'd started this effort to establish a standard to help my *siblings* discover their true genetic heritage, and to establish health histories for all of us, if needed. Now Jack was telling me that it was no longer just for my siblings, but for me and my own posterity as well.

I just couldn't accept this result. I *knew* I was Mark's child! I asked Jack if the fact that the samples had been kept in the freezer for so long could have made any difference. He assured me that the samples were fine. I asked him if there was any way we could do a second test on the samples. I just knew there had to be a problem somewhere in this process. This just wasn't possible.

Jack asked if we had any other DNA samples from my dad. I reminded him that the samples we had obtained were from a dead body lying in a coffin, and that the body was now buried six feet underground. We were not about to request an exhumation. When I mentioned this to Jack, I was not only telling him that it was impossible to get any further samples from my dad's body—I was also reminding him of Dad's condition when we took the original samples. I was hoping that Jack would tell me there was a chance that the embalming process might have contaminated the hair samples. But this was not the case. He reminded me that after a person is dead, their body is no longer able to take nutrients or other chemicals into their hair or fingernails. This meant that none of the chemicals used in the embalming process would have had a chance to get into his hair.

Jack asked what kind of razor my dad used. I remembered that Dad always preferred an electric shaver over a straight-edge razor. I also recalled Tim telling me that when our sisters were cleaning out Dad's stuff after his death, they had decided that Tim should have Dad's electric shaver. Tim said that when he pulled it out of its carrying case, he noticed that the hair chamber was filled with stubble. He had put it away, never giving it a second thought. This electric shaver hadn't been out of its case since Tim had gotten it more than two years ago.

I told Jack about the shaver, and that it had never been used by anyone else other than Dad. Jack was excited to hear it. I reminded him that there wouldn't be any hair follicles on the stubble, since any whiskers in the hair chamber would have been cut off at skin level. Jack helped me to understand that it wasn't the hair he was interested in. What he wanted were the flecks of skin that always come off the face when someone shaves, even with an electric shaver. He was sure that the chamber would have a rich supply of skin flecks, along with all the hair stubble. I told him that I'd call Tim and have him send all the hair in the chamber to Jack's office so he could run a second test on the sample. By now I didn't care how much it cost; I just wanted to know for sure whether or not I was Mark Anderson's son.

Looking back at this turning point in my life, I am fascinated by the incredible flood of thoughts and emotions that filled my mind. In fact, my whole body was shaking. It surprised me, how strongly I reacted to this news. I'd never experienced anything like this before. It was like there were a thousand different emotions, all vying for my immediate attention. My mind was on overload in just a fraction of a second. It was quite a remarkable sensation. I'm not sure I liked it.

I had always been very close to both my mom and my dad. Heaven knows, my mom had plenty of faults; some of them were serious. She had hurt a lot of people. Everyone in the world has some bad traits, and our mom had more than her fair share. But I must say that she was good to me, probably more so than she was to most of my siblings. In fact, all of my siblings continue to remind me that Mom always liked me best. I'm not sure if that's true, but looking back on how things went, I suppose they may be right. We were closer than most.

Tim and I were especially close to Dad. I think he realized that he'd been a bit hard on the older kids, so he made an extra effort to be a little more patient and protective with us last three kids. I don't have a lot of memories of Dad getting angry at me. I think I can remember being spanked by him just once, even though I gave him plenty of reasons to spank me more times than I care to think about.

Now I was learning that Mark wasn't really my father, and that Mom was getting frisky with more guys in town than I'd originally thought. This was more than I could handle at the time.

After I got off the phone with Jack, I just stood there, alone, in shock. I'd been at work when he called. Given the sensitive nature of the call, I'd decided to go outside to the parking lot for our conversation. Now here I was, standing in the lot, unable to move. I just stood there, completely still, wondering what had just happened. I had just been blindsided by news that I was totally unprepared for. I was stunned, and there was nothing I could do to help me deal with what I had just learned. It was like time was standing still, and I couldn't for the life of me figure out how to get it moving again.

When I finally got to the point where I could move again, I couldn't stop. I walked back and forth through the rows of cars. I simply didn't know what else to do. I could put one foot in front of the other, but that was about it. I had to keep walking, because as long as I was moving, I had at least some sense of control over who I was. But then I realized that with this one phone call, this was no longer true. Science and technology had stripped me of whatever sense of identity I'd thought I had.

In a matter of minutes, my entire identity had changed. I was no longer the Steve Anderson I'd thought I was for the past fifty-eight years. I wasn't even an Anderson. What made matters even worse was that I had no clue *who* I was. If I wasn't an Anderson—if Mark Anderson was not my biological father—then who was I? I was going through a classic identity crisis. I felt like I was going to throw up, so I stayed outside and continued to walk through the parking lot. I can't help but wonder what my coworkers must have thought when they saw me walking around the parking lot like an old homeless man with no place to go.

I'm sure this must have been what my brother Neil went through when he found out that Mark was not his father. He had it even worse than me, as he was only sixteen when he found out. Those teen years are hard enough for any kid, but to be told that the dad you grew up with and loved wasn't really your dad, and to think you were the only kid in the family going through this—it must have been devastating for him. At least I was a fully grown adult and could share the experience with three other siblings who were confident that they had a different birth father. This gave me a profound sense of respect for my brother Neil, considering all of the emotional garbage he has had to deal with because of what he had learned at such a young age. It's no wonder he's never been very close to our mom.

After pacing through the parking lot for what seemed like forever, I felt an enormous feeling of rage growing in my chest. My thoughts were racing, and I started talking to myself, something I don't often do. I remember saying out loud, "I can't believe my mother would do something like this!" A sense of utter betrayal clouded my mind, filling it with all kinds of confusing questions and thoughts. I was an emotional mess!

I thought about the small town I'd grown up in, one of those suburbs outside of Chicago with a total population of less than five thousand people back in the late 1950s and early 1960s. I knew nearly everyone who lived there, and they all knew me and my family. It suddenly dawned on me that many of these people must have known some of this story that I was only beginning to learn. In a small town like ours, nothing escapes the prying eyes of the town gossips, and we had plenty who filled that role. I remember some of my school friends telling me that they were not allowed to come and play at my house. I never knew why their parents wouldn't let them come over. Now it was starting to make sense: Their parents probably didn't want them spending time at my house because they must have known about my mother's exploits! I was mortified. If I'd known this at the time, I would have been so humiliated that I never would have come out of our house. I was terribly shy and socially awkward as it was; something like this would have caused me to withdraw from society altogether and hide myself in some kind of womb of my own making.

I kept wondering why Mom would do this. Then I started feeling angry at Dad for letting this happen. Why didn't he do something to stop Mom? I was child number seven. Surely he must have known what was going on by the time I arrived. The only other alternative was that Mom must have been amazing at keeping secrets, but I couldn't believe this was even a remote possibility. Mom couldn't keep a secret if her life depended on it. Surely Dad must have known what was going on. Why didn't he say something about this to us? But no; in fact, he seemed to go out of his way to protect Mom. I remember once being mad at her for something and making an angry comment about her to my dad. He immediately told me to never talk that way about her again; he said that she was my mother, and I should always respect her as such.

I couldn't believe the feelings of betrayal and rage I was experiencing. I don't think I had ever felt anything like this before. It felt like my whole world was coming down around me. I'm not one who normally gives in to drama. In fact, I find it a bit silly to watch people who live their life going from one dramatic episode to another. But now I was going through some serious drama of my own, trying to come to terms with very real emotions. I knew I would never be able to focus at work, so I left and went home.

Below is a journal entry I wrote at the end of this day. I think it captures how I felt at the moment I learned all of this life-changing information:

> August 22, 2012
>
> Well, I was in for quite a shock today. I spoke with Jack Anderson, the guy in Oceanside, California, who is doing my DNA testing. He called to tell me that the tests were done and that he's sending me the results. When I was there with Tim a couple weeks ago, Jack showed me the report I would receive after the testing was done. It was all very, very complicated, so I called him to see if he would just tell me what the results were, and I could ask him any questions I might have.
>
> Jack told me that 16 factors must be positive for paternal parentage to be established. In other words, if all 16 factors are positive, then the person in question is the real genetic father. One of those 16 factors coming up negative would bring that parentage into question. In my case, only 5 of the 16 factors were positive. There were 2 that were questionable, and 9 were

negative. In other words, Dad is definitely NOT my biological father.

I couldn't believe my ears when I heard that. I was in shock. Of all the kids in our family, everyone was sure that I was an Anderson. I look like all our uncles and many of our Anderson cousins. I never for a moment thought it could be any other way.

I was surprised by the range of emotions I struggled with for hours after the phone call. I was so angry at Mom for what she had done. She had betrayed Dad not just once, but at least four times (keep in mind that we know for sure Neil isn't Mark Anderson's son, and we're fairly confident that neither Holly nor Carlee are Dad's biological daughters). Was Mom sleeping with every sex-driven man in town? Is she some sort of floozy who would let any man jump in the sack with her?

Then I was angry at myself for feeling so angry at Mom and for being such a hypocrite. I remember telling Jack that if I found out Dad wasn't my biological father, I wouldn't be too shaken by it. I thought I would handle it just fine and take it all in stride. Boy, was I wrong. I feel like such a fool.

This leaves so many things unanswered for me now.

- If Dad isn't my dad, who is?
- Will Mom even remember who my biological father is?
- Is there any chance that he is even alive?
- If he is, is it appropriate for me to let him know that I'm his son, after all these years?
- What would such a revelation do to him and his family?
- What about all the family history work I did for the Anderson line?
- How about all those feelings of closeness I have felt for Grandma Anderson, Jacob O'Neil, and the others? They aren't even my blood family now.
- How many of my other siblings have been fathered by other men?
- What is wrong with Mom that she was such a loose and promiscuous woman?
- What genetic issues and problems do I need to be aware of now that I belong to a different genetic line? I used to be extra

sensitive to the possibility of diabetes and cancer, since it is such a prevalent trait on my Anderson line. What new health issues do I need to be mindful of now that I know I'm not an Anderson?

- What will my kids think knowing that at this point, they have no idea who their paternal grandfather is, and that their grandma was a . . . well, we won't go there, at least not now.

I have so many questions that are running through my mind as I try to come to terms with this. I feel like I am on an emotional roller coaster, and I feel like I'm going to explode!!! I find myself looking at my own skin, my arms and hands, this face I see when I look into the mirror, and I wonder who contributed to making this body? Whose genes am I carrying? Is my biological father dead, or is he still alive? Will I ever know who my biological father was? Mom has lied to me so many times already that I wonder if I can even trust anything she might tell me about this, or anything else.

I never realized why Neil felt so separated from the family. I understand now. I don't think I would ever push myself away from the family like Neil has, but I certainly can understand why he feels the way he does. I am shocked that I feel this way. Part of me feels ashamed that I am reacting like this, while another part of me feels so betrayed and angry at Mom.

I asked Jack Anderson to see what more we can do to remove any possible doubt that there could be a mistake in his tests. After all, the DNA sample from Dad is two years old, and he said that he didn't find any hair follicles in the hair samples Tim provided. Fortunately, Tim was given Dad's electric shaver when Dad died. He put it away without even taking it out of the holder or touching it. Jack said that it's not the hair he needs but the microscopic skin flecks that will be in there with his hair. I decided to pay and have him go ahead and run additional tests on whatever he can retrieve from the shaver. It's going to cost me to run another test, but if the rest of the kids in the family are going to want to have their DNA tested, they certainly need a reliable set of genetic data to work from on both Mom and Dad's side.

> Ohhhh, my head hurts. The muscles in my chest are so tight. I'm having a hard time thinking straight. This is such a shock to me. I don't even know how I feel about this.

When I got home, my wife could see immediately that something was terribly wrong. I told her that I got the DNA test results back from Andergene Labs and explained what Jack had told me. The only thing she could say was, "Wow! That's impossible!" I didn't realize it at the time, but my twenty-five-year-old son was in the kitchen and overheard me talking to my wife about it. When he came walking into the room, the first thing he said was, "Wow, Dad, how does it feel to know that you're a bastard?" As soon as he said it, he realized that this was probably not the right thing to say, and it certainly wasn't the best time to bring it to my attention. But he was right. I was a bastard!

Once again, a flood of emotions raged through me. I was at a loss as to what I could do about them. I was confused. I was so very angry. I knew that I needed to show a degree of respect for my mother simply because she was my mother. She had done some awesome things that changed my life for good in some very important ways. Yet this new revelation blinded me to anything good she had done for me. I only felt feelings of contempt for her now, and I hated myself for feeling that way. Those feelings only added more fuel to the fire that was already burning so fiercely inside me.

I had to call Tim and share this discovery with him. Tim and I had always been close. Since we were little, we had shared everything. We ran track and cross-country together, and shared school assignments. Most of my friends were Tim's friends. We did things together all the time. In fact, I was held back in second grade because I couldn't read well enough to keep up with my schoolwork. It was decided that since I was being held back, Tim would be held back as well. So, for better or worse, we even shared our grade school shame. Now, I needed to talk with him. I wanted to see if he could help me come to terms with what I was feeling.

When I told Tim what Jack had told me, he had the same response as my wife: "Wow! I can't believe it! That can't be!" After a couple of hours of talking with him, I calmed down, and we started thinking about the implications of this new revelation. Over the past several years, we had thought about the likelihood that Neil and two of our sisters were not Mark's children. We had both thought, wouldn't it be funny if we found out that Dad was sterile? I was beginning to wonder if there might be some truth to that possibility.

By the end of our discussion, we decided that with this latest revelation, maybe we needed to get Tim tested to see if he was Mark's child. While I was sure that Tim was Mark's son, I would have bet everything that *I* was Mark's son, too, and we knew how that had turned out. Tim had thought the world of Dad; they were very close. It was no secret in our family that Tim had felt a lot closer to Dad than he felt to Mom.

The next day, Tim visited Jack at his genetics lab and delivered the shaving residue from Dad's old electric shaver, which Jack needed in order to do a second test on Dad's DNA. Tim also asked Jack to run a DNA test on him. Jack was more than happy to run a paternity test on Tim, so Tim provided the routine cheek swab samples and a check to pay the costs of testing.

Now it was time to wait. The waiting has been the hardest part of this whole ordeal.

Jack knew I was struggling with what I had learned from the first test he'd done on Dad's DNA, so he made the second test a high priority. When he had the results, he called me and told me they were the same as the results from the first test. The skin flecks Jack had retrieved from the shaver indicated that Dad's DNA did not match mine. I was not Mark Anderson's son.

Now there was no denying the truth. There was no reason for me to hold on to the hope that some kind of terrible mistake had been made. I was not Dad's son, and nothing was going to change that. It was time for me to start coming to terms with the truth.

A month later, I got a call from Tim, who didn't sound like his normal self. His message was short and to the point. He simply said, "I got the test results from Jack. Dad is *not* my biological father."

What! Are you serious?! Once again, there were feelings of shock, this time from both of us. My first thought was, "How far down the rabbit hole does this story go?" Little did I realize that it would go a lot further before we would begin to see light at the end of the tunnel.

After hearing that Tim was also not Mark's son, I decided then and there that I needed to buy a plane ticket home and confront our mother. She was the only one who might have the answers to all the questions we had. I wanted to see what she had to say about all of this. We needed some answers, and confronting her was the only way we were going to get them. The answers were in our hometown, where Mom still lived, and I had to go there to find them.

I had no idea at the time that I would find a lot more than I had bargained for. And the more I learned, the more questions I would have.

5

FINDING THE ANSWERS

I love going back home to be with family and friends. However, planning a trip home to tell my mom that I'd discovered I was not fathered by the man I'd always considered my dad made this trip anything but something to look forward to. I was trying to figure out what I was going to say to Mom—how I was going to tell her that I knew her secrets. I think I was also worried about what other secrets I might uncover during this trip. I could feel my stomach tighten up every time I thought about seeing Mom again.

Mom was ninety years old when I flew back to visit with her. Her body was worn out, and she could no longer take care of herself. Just a few years earlier, we had had to move her into a full-care facility. However, despite her feeble condition, her mind was sharp. I was always amazed at what she could remember; there was no pulling anything over on her.

My brothers and sisters were surprised to learn that I was coming back home. For the past several years, about the only times I came home was for weddings and funerals, or the big mile-marker birthdays, when someone turned seventy, eighty, or ninety.

My oldest sister, Holly, had already passed away by the time I made this trip. I only wish I could have found out what she knew before she died. I think that if she would have known what I had discovered, she would have given in and told me all the stories she had been hiding for years. She knew a lot about the family, and I had wanted to get her perspective on many of the conflicting stories that had been passed around. The early years in our family were so different from the later years, when Tim and Carlee and I grew up. I kept hearing about events and people I knew nothing about. More than a decade separated me from the older sisters. I was interested in learning more about what life was like in those years, but Holly had never been willing to help me find what I was looking for.

These days, I think I agree with Holly that some family secrets do need to die; many people would probably agree. Sometimes it's not worth dredging up a painful past. Some secrets have the power to do enormous damage, and uncovering them would do no good from generation to generation. On the other hand, it can be a tragedy for some family secrets to experience a premature death. Some are so powerful that, as painful as they might be, they have something enormously valuable to offer and need to be revealed. I believe without a doubt that the knowledge of one's genetic heritage is one of those things that must never be hidden. Like it or not, much of what we are is embedded in that genetic code that makes us who we are, and we cannot afford to lose that information.

I arrived home late, so I called Mom and told her I would be over to see her early the next day. I had recorded part of her oral history over the telephone a few years before, so I told her that the purpose of my visit was to get more of her history and ask her questions about some of our immediate family's history. I said I'd come over to record more of her memories, assuring her that it would be a very exciting interview. She agreed, although she had no idea just how exciting it was going to be.

That night, my stomach was churning, and I spent hours trying to figure out how I was going to bring up the subject of what I had discovered. This was not going to be easy. Try as I might, I couldn't seem to find any words to use that wouldn't sound like I was accusing her of something terrible. That was not my intention at all. I just wanted some answers so I could make sense of my new identity. I wanted to know who my father was, and understand why things had happened the way they did.

MY SEARCH FOR THE TRUTH

I arrived bright and early the next day. Breakfast had already been served, and Mom was back in her room, waiting for me. We worked on her oral history for about thirty minutes while I mustered the courage to confront her. I knew this was not going to be easy.

Mom and I had always been very close. In retrospect, I think she believed that if I ever discovered the secrets of her infidelity, she would probably lose me. She had lost Neil because of this very thing and couldn't bear the thought of losing another child this way. Yes, I had been shocked to learn that I was conceived during a tryst with another man, and for a while, I was very angry with her. But with time, I had started to heal, and I was pretty sure that she would not lose me. I had to convince her of this so that she would feel safe enough to tell me the truth—all of it.

Finally, after half an hour of sharing stories, I decided I just had to come right out and tell her what I had learned, and ask her point blank who my real father was. I started my conversation something like this: "Hey, Mom, Tim has a friend who owns a genetics research company. I had my DNA tested to see what it would tell me. You can find out all kinds of things with these tests. You know what I learned? I found out that Dad isn't really my genetic father. What can you tell me about that?" Simple and to the point. I tried my best to not sound combative or accusatory.

Without missing a beat, Mom said, "Oh, you can't really trust those DNA tests. They aren't all that accurate, and the results are just wrong a lot of the time."

I told her that initially, I'd thought the same thing, so just in case there could have been an issue, like the sample being bad, or the test results being incorrect, I had used a totally different sample for a second test, just to be sure. I told her the results of the second test were exactly the same.

It was quiet for a minute. I could tell she was thinking about what to say next. I was sure she would crack and tell me who my father was. I didn't want to overwhelm her with the news that we'd had Tim tested, too, and that his results also revealed that Mark was not his father. I'd tell her that later, after she'd gotten over the shock of learning that the jig was up, and we were on to her secrets.

Instead of telling me the truth, Mom simply repeated that the DNA test was wrong, and that Mark Anderson was indeed my father. She then told me she didn't want to go on with this conversation any further, and that we needed to get back to talking about her history.

I was completely at a loss for words. She was lying to me, and she made no apologies about it. If I pushed back, I knew she would clam up, and I'd never get any information about who my real father was. I didn't know if Mom understood the situation we had here, but I knew that if she decided not to talk about this now, then none of us would ever know anything more about who our fathers were. We would never have any idea who really belonged to Dad and who didn't. I wasn't ready to take that chance.

This was a very tough position to be in. I had recently read a book titled *The Stranger in My Genes*, written by a man named Bill Griffeth.[1] He also discovered, through the use of a commercial DNA test, that his dad was not his real father. He was visiting his elderly mother to see if she would be willing to provide some additional information about his biological father and the circumstances of his conception. His encounter ended up very much like mine did. She simply couldn't, or wouldn't, talk about it. Situations like this are a delicate balance between someone who is desperate to find out who they are and how they came to be and someone who feels like they've made a terrible mistake and have lived with their shame for many years. Circumstances like this are never black and white, and have the potential of doing great damage to a relationship if not handled right.

I realized that confrontation at this moment would do more harm than good. I could feel myself getting angry that she wouldn't own up to her involvement in such a critically important matter. I told Mom that I had something I'd promised I would do for my sister Diane, and that I needed to go. I grabbed my stuff and was gone from her room in a matter of minutes. I usually give Mom a kiss and a hug when I leave, but in this case, I just couldn't bring myself to show her any kind of affection. I had to leave right away, before I said something I would regret later.

I got into my car and just sat there. I was furious and didn't know what to do. I typically don't get truly angry; I'm usually an easygoing guy. In fact, I can remember only two times in my life when I was ever filled with rage. The first time was when I was eleven years old and got into a big fight with my older brother for destroying some trees I had planted. The second time was when I got the DNA results back from Jack and discovered that Mark was not my biological father.

Now, I was feeling those same feelings of rage again, and I didn't like what it was doing to me. I wasn't familiar with this level of anger, and have to admit that I was quite surprised by the enormous power it has on the soul. I knew I had to do something, and quickly, to get these feelings to stop.

People do odd things when they get really angry. When I get mad or frustrated, I walk. I walk hard, and fast, to work out the raw feelings. I knew that I had to walk right now. Within a stone's throw of the care center where my mom lived was a cemetery. I've always thought it odd to locate an end-of-life care center so close to a cemetery. It seems a bit grim to be able to look out your window and see where your body is going to be deposited when you finally leave this world. In a matter of minutes, I was in the cemetery, walking my heart out. I'm not sure why I chose to go there; maybe because it was quiet, with no one to bother me. I walked over to my grandparents' tombstones and just stood there, wondering what in the world had happened to their daughter, that she would be messed up enough to have all of these kids out of wedlock—not just one, but five. Maybe all nine of us were conceived out of wedlock! For all I knew at this point, that could turn out to be the case.

A strong winter wind was blowing as I walked through the rows of tombstones. I was so cold that I finally had to go back to my car and warm up my hands and my ears. Once my lips warmed up enough that I could talk again, I called my brother Tim and told him how things had gone with Mom. I guess he hadn't expected it to go well from the beginning, because he didn't seem at all surprised when I told him that Mom wasn't willing to talk about the matter, or to give me any information about what really happened. I am always amazed by Tim. When things bother him, he doesn't seem to lose his temper or go crazy. He just handles it with a calm sense of dignity. He is one of those people who makes you feel a bit silly for getting all worked up when you know it's going to be okay in the end. He was handling this a lot better than I was.

We talked about what had happened and what our next moves were going to be. I was hoping he had a strategy that I could use, because I had no idea what to do next. He assured me that somehow things were going to work out, and that we'd discover what we needed to know. After about an hour of talking with Tim, I was feeling much better.

I called Carlee next. As I told her what took place, I could feel myself getting all worked up again. Carlee calmly guided me back into the real and rational world again, and helped me see things from a better perspective. After talking with her, I decided I'd walk around the cemetery again and see if I could figure out what my next move was going to be.

I went to Dad's grave to see if he had any ideas. Nothing. I knew he'd probably had all the answers I was looking for, plus a few more I hadn't even considered yet. I know it wasn't possible for Dad to send me answers from beyond the grave, but at this point, I was willing to entertain the thought that just maybe he would come back for a visit and tell me what I wanted to know. Then I wouldn't have to try and get answers from my mom anymore. No such luck. It's funny to see what lengths we'll go to when we get desperate.

After spending more than an hour in the cemetery, I went back to my brother Neil's home where I was staying during my visit. I had to smile as I thought about how I'd chosen to stay with him during this critically important trip. It was his tractor accident that had started this whole adventure.

When I got to Neil's house, I told him the real reason I'd come home for a visit. I didn't tell him all the details that had gotten me to this point—only that I had found out that Mark wasn't my dad. He didn't seem at all surprised. Things had happened between Neil and

Mom that made a close mother–son relationship impossible for either of them. I don't know the specifics of all the conflicts, but I knew enough about some of the events that it was plain to me she was punishing him for something that none of us knew about.

I can remember one event in particular that stood out as being especially unfair to Neil. He has never forgiven Mom for what she did to him, and Mom knew that.

When Neil was about sixteen, shortly before his tractor accident, Dad brought home a couple of piglets for him to raise. At that time, we lived on what used to be an old farmstead complete with a barn and a chicken house. Neil fixed up a section of the barn so that it would accommodate the two young piglets. He figured he could raise them until they got full size, and then he would sell them and make some seriously big bucks. He had his driver's permit and knew that he would be getting his license soon. With the money Neil would earn from raising his two piglets, he could pay for his own car.

About that time, Dad's aunt and uncle were visiting from Oregon. While they were at our house, they invited Mom to come to Oregon with them and spend some time getting to know the rest of Dad's aunts and uncles. While she was there, she could visit Grandma Anderson and have a lot of fun doing things together. The only problem was, Mom didn't have any money to pay for the trip. Somehow, she scraped up enough money to get there, but didn't have enough to get back home. Mom's solution? "I'll worry about getting back when it's time to come home." She was never very good at working out the finer details of anything that involved money. She just went with what was happening at the moment and figured she'd worry about tomorrow when tomorrow came. While this rarely worked out well for her, she went through most of her life living by that motto.

After staying in Oregon for three weeks, she called and told Dad she needed money to get home. Dad told her that he didn't have the money, but he'd ask around and see what he could come up with to buy a bus ticket back home. But Mom didn't want to travel by bus; it would take too long, and plus, it wasn't much fun. She wanted to fly home. Dad told her that he couldn't come up with enough money for that; she'd have to stay there for another month before that would happen. Mom told Dad he could get all the money she needed to buy a plane ticket if he'd just sell Neil's two pigs and take whatever he could get for them. There was no time to haggle for a good price. She wanted to get back home, now. So Dad sold Neil's pigs and bought her the plane ticket. Mom was happy to get home, and quite enjoyed being able to fly in an airplane. She had never flown before, so it was quite the adventure for her.

Neil was not at all happy with what had happened to his pigs. As far as he was concerned, Dad should have let Mom stay in Oregon and find a job to earn the money she needed to get back home. Over the ensuing years, Mom and Neil had many fights over her "theft of his pig money." He has never been able to forgive her for selling his pigs. He knew that Dad wasn't really to blame for that. In his eyes, it was Mom who was responsible for the loss of his pigs, which ultimately meant the loss of the car he planned to buy with the proceeds.

So, when I got back from the cemetery and told Neil that I had found out Mark wasn't my father, he just shrugged his shoulders and said, "Well, I guess that doesn't surprise me one bit." That wasn't quite what I was expecting, but at least he accepted the news better than I thought he would.

The next morning, I got a call from Mom asking me to come back and visit with her. She said we needed to talk. This came as quite a surprise to me, but I was glad to hear that maybe I had a second chance at getting the information I needed. I went over to the care center after I knew she would be done with breakfast and back in her room.

She told me that she had spent a long, sleepless night thinking about what I had said, and how she was going to tell me what I needed to hear. She came right out and told me that the DNA test was correct: Mark was not my biological father. I asked her who my biological father was, and she told me it was a man named Ray Jacobson. I thought about it a minute and couldn't recall anyone in town by that name. The only Jacobson I could remember was a girl in my class at school, named Diane Jacobson. Then it hit me: Was she my half-sister? I knew Diane, and recalled that I didn't much care for her. She was one of those high-performing kids that always looked down on kids like me.

I was never one of the gifted kids in school. I couldn't read very well when I was in second grade, so I struggled to make good grades. I remember that our teacher had the kids in our class separated into three different reading groups. Those who read above average were in the Robin reading group; the average readers were in the Bluebird group; and those who struggled were in the Blackbird group. I spent most of my time in the latter. The teacher wasn't very encouraging to our group, and seemed to spend considerably less time with us than she did with the other two groups. This humiliating experience was one of the lowest points of my early school years. Because of this, I wasn't especially fond of kids like Diane Jacobson, who were in the Robin reading group. They treated me like I was stupid, which is probably the way I acted much of the time. I didn't handle the idea of being in the Blackbird reading group very well and often acted out because I felt humiliated—at least, that's what my teachers told my parents. Either way, Diane was a "Robin reader," and because of that, I didn't much care for her.

Then Mom added something that really muddied the waters for me. She was thinking out loud and said, "Yes, I think it may have been Ray Jacobson. It was either Ray, or it was his brother, Timmy." *What?* Brothers? Wow, that's convenient. She'd been seeing both brothers? This was getting more complicated than ever. It looked like I was going to have to do further research to see if I could determine which of the two brothers was really my dad.

Of course, there was always the possibility that Mom wasn't remembering the details correctly. If she couldn't remember which of the two brothers was my father, then it was only logical that this might not have been an isolated incident, and that she was probably pretty promiscuous in her younger years. At ninety, she might not even remember who my biological father was. I mean, nearly sixty years had passed since all of this took place, so I wasn't sure if I could trust her memory.

I decided that as long as she was spilling the beans about who my father was, now might be the perfect time to ask her about the letter my siblings and I had often talked about. None of us knew for sure if this letter really existed, but now was the time to ask.

I told Mom that a story had circulated for decades among us kids about a letter she had supposedly written when she was in her forties. During that time, Mom had thought she was dying from ovarian cancer. The story was that before she died, she wanted to write a letter to her children revealing the truth about a lot of things that I had never heard about. I was told that Mom gave it to her sister Eva to keep, asking Eva to give the letter to our oldest sister, Holly, after Mom died. I explained to Mom that my older sisters had sworn such a letter existed, and that it contained all the secrets she had wanted to confess before she died. I came right out and asked if such a letter really did exist. I had nothing to lose and a lot to gain if she confirmed its existence, and if she'd tell me who might have the letter now.

Mom seemed surprised that I had heard about this and denied the letter's existence. If such a letter did exist, she said, she wouldn't have given it to her sister Eva. Eva would have blabbed everything all over town. I pressed her a bit more to see if I could discern whether she actually knew of it and was denying its existence, or if maybe she simply didn't remember it. After a few minutes, I concluded that if such a letter really did exist, she wasn't going to tell me about it. I was walking up a dead-end street with this issue and could see it was time to move on to other questions.

I decided that if I was pushing her for information and she wasn't getting angry at me for being too nosy, I would go for the big prize and see what I came up with. I reminded her that some of the kids and grandkids were dealing with health issues; I talked about the great new advances scientists were making in medicine because of genetic engineering. Then I reminded her that if we didn't have the right information about our genetic past, that someone could die, either by getting bad information or by missing out on some treatment opportunities that might be available to us if we knew what illnesses were showing up in our family lines.

After that, I came right out and asked her who the fathers of each one of us kids were. I started with the three we already knew about. First was Holly. Mom told me that she was Sam's daughter. Then I asked about Neil. Oh, he was Dennis's son. Carlee was Peter's daughter. Great! I was batting 100 percent so far, so I was ready to go all the way and ask about the rest of us.

With the name of each man she gave me, she provided a bit of a biographical sketch, which I thought was interesting; it was certainly more than I'd expected at this point. With some of the men, her memories were sharp and very detailed. With a few of them, I could hear in the inflections of her voice and the moments of silence that she was feeling a sense of longing—like she sorely missed being back in those times, once again enjoying the company of the man she was talking about. It was very clear that with some of these men, she'd had longtime relationships that meant a great deal to her. I would have given anything to have her put aside all of her inhibitions and tell me more about these relationships. I'm not talking about the seedy, intimate physical details; I'm talking about how the relationships developed, the ebb and flow of their experiences together, and how serious each romance was to her at the time. I am always fascinated by people's love stories. With my mom's multiple experiences, I was especially interested in learning why things turned out the way they did, and why she'd felt the need to look outside her own marriage relationship to fulfill her romantic needs.

She had already told me that Ray Jacobson was my biological father—well, either Ray or his brother, Timmy. I would ask for more details about this man later on, but for now, while she didn't seem tired and I was on a roll, I wanted to learn more about my siblings' fathers.

When I asked her who Tim's father was, Mom piped up right away and told me it was Timmy Jacobson. Wait a minute! She had just told me that Tim Jacobson could possibly be my father. While she wasn't sure which of the two Jacobson brothers was my dad, she was positive that Timmy was Tim's dad. Almost immediately, she went back into her reflective mode and started talking about how handsome Timmy Jacobson was. She said that every girl in town wanted to be with Timmy. In fact, she'd named my brother Tim after Timmy Jacobson. Hmm, I thought; that was a bit cheeky, to name a son after the man who had fathered him, when the guy wasn't her lawfully wedded husband—but at this point, I could expect almost anything. I could understand if this guy had been having an affair with a single woman, but in this case, where Mom was married and had a family with seven kids already—well, that's just not right.

I really wanted to know what had been happening in her relationship with Dad that would have led to her having all of these extramarital relationships. And why was Timmy Jacobson, who was single and about twenty-eight at the time, fooling around with a married woman? This was getting crazier with each question I asked. As weird as it was to hear about Timmy Jacobson, I was delighted to know that at the very least, my brother Tim and I both came from the same paternal family line. From what I could tell so far, we were a bit more than half-brothers.

By this point, I knew the names of the fathers of five of us nine children—at least, according to my ninety-year-old mother. Not bad. Now I asked about Diane.

Mom told me that Diane's father was a guy named David Smith. When I first heard that, I had the distinct impression this was bogus; it sounded like a very generic name someone would use to hide their real identity. Now I was confused. If Mom was being truthful about the others, why would she lie about Diane's father? I knew most of the people in town, and I didn't remember a man by the name of David Smith. I knew something wasn't right with this bit of information.

So now, I was six for nine, with some uncertainty about number six. When I asked Mom who the fathers were for the others, Mom assured me that Paul, Gloria, and Judy were all Mark's kids, without a doubt. All three had been fathered by Mark Anderson, the man who raised us. She was sure of that and didn't offer anything more about these last three names.

I had all nine kids accounted for now. However, just like Mom's answer for Diane, something didn't feel right about her comment about Paul, Gloria, and Judy all being fathered by Mark Anderson. I don't know if it was her body language or the way she said it, but it just didn't feel like she was telling me the truth about the fathers of Diane, Paul, Gloria, and Judy. Was she lying, or was she not able to remember? I had a hard time believing it was a memory issue, as she'd been very straightforward about telling me the names of the fathers for five of us kids. For some reason, she seemed to be hiding something about the last four children. I had nothing of any substance to prove she wasn't telling me the truth; I was just going on a gut feeling.

After we talked a bit more, I decided it was time to go back to Neil's and mull everything over. As I was leaving her room, Mom said, "Steve, do you hate me for what I've done?"

I assured her that I didn't hate her. But I was also honest, saying that all of this was so weird and crazy, I would need some time to figure out how it all fit together. I wasn't sure how I felt about any of this.

She laughed and agreed. She reminded me that she'd never made life easy for anyone. No apology; that's just how it was.

I would have loved to have taken the time to probe into that comment a bit more. Through all of this, I kept asking myself what had happened to her that she would have had so many extramarital relationships. According to Mom, there were at least six relationships that ended up in pregnancies, and possibly all nine. I wondered if there had been many others who'd come and gone without her getting pregnant. I had to wonder if there was some kind of arrangement between Mom and Dad that had allowed each of them to have relationships with anyone they wanted. Did they have an open marriage? Was that okay with both of them? I just couldn't picture my parents as swingers; they didn't fit that image. Did my mother experience some kind of abuse as a young girl? Did she have a sexual drive that made her willing to do almost anything to get the love and approval of any male figure willing to give it to her?

My parents would eventually divorce in 1965, after twenty-five years of marriage. I don't know all the reasons for their divorce, but surely this infidelity of one—or both—of them must have played a role.

I felt lucky to have obtained the information she'd given me that morning. I had come to her room doubting that she would be willing to discuss any questions that dealt with such personal issues. But I must say, I would have given anything to be a fly on the wall, listening to a conversation between her and a psychiatrist. I would love to know what was going through her head.

Before I left she made me promise that I would not talk about any of this with my sister Diane. My sisters Diane and Gloria were two of the few people who ever came to visit Mom on a regular basis. She knew that if Diane found out about this, she would be furious, and would probably never come to visit her again. If she did come to visit, it would be only to badger her into telling her more. She also knew that Diane would likely go around town telling people about what she had discovered, and asking people who her father might be.

I called Tim to let him know the news about Timmy Jacobson. He was interested, but didn't seem shocked by the news. He was pleased to hear that it was likely our fathers were brothers. We'd been close all our lives; we seemed to share a bond that was stronger than most. News about Timmy and Ray Jacobson just helped to reinforce the sense of brotherhood that Tim and I had always enjoyed.

I met with Mom a few more times to finish completing her oral history. However, I finally gave up on that. I knew that too much of what she was telling me was not true. Mom had a way of changing her perceptions of things to fit her needs. If anything was unpleasant, reflected badly on her, or didn't fit the image of what she wanted her reality to be, she simply changed the facts. Once she had introduced a fabricated memory into her mind, it became truth to her. Listening to the things she told me, I was amazed by her remarkable imagination. While it made for a great story, as far as being a reasonably accurate personal history, it failed miserably.

What I had recorded of her personal history was of no real substantive value whatsoever.

When I got back to Neil's house, I called my sister Gloria and told her what Mom had shared with me. She knew why I had come to visit Mom and didn't seem too surprised that Mark was not my father. When I originally told her that I was going home to talk with Mom, I'd promised that I would report back to her what I had found out. Gloria has been my sounding board almost since the beginning of this journey, and has kept me from getting sidetracked as I look for answers to my many questions. She keeps me focused and prevents me from giving up in despair.

Even though Mom had told me that Paul, Gloria, and Judy were Mark Anderson's children, I was not convinced she was telling the truth. Neither was Gloria. For years, the older girls had joked about Paul not being Dad's son. Paul's name was Paul Michael Anderson, and Mom had often entertained a guy by the name of Michael Paul Keller. Gloria said that he was around the house often enough that the girls wondered if their brother was really Keller's son, and not Dad's. Paul never looked like an Anderson, but he did look amazingly like Michael Paul Keller. With this in mind, Gloria wasn't convinced that Mom was telling the truth, either—at least where Paul was concerned.

With the success of Tim's and my DNA testing experiences, Gloria decided she would like to have her DNA tested, as well. I'm not sure if she seriously questioned her paternity or just wanted to be 100 percent sure, but she decided she had better be tested so she could know the truth. Tim sent her a swab kit with instructions on what to do. In fact, Tim decided to pick up several kits. He knew chances were good that as we continued to investigate this mystery of our family's heritage, one or two kits would most likely not be enough; we'd probably need several.

GLORIA DISCOVERS THE TRUTH

Gloria got the results back from Andergene Labs about two months later and discovered that she, too, was not Mark's daughter. Although she'd had her doubts, at the same time, she was surprised to learn for sure that this was the case when the test results came in.

I talked to Gloria about this latest discovery and what it meant to her. I asked what her next step would be. Of course, she wanted to know who her biological father was, but the thought of asking Mom seriously scared her. Gloria was from that generation where you just didn't talk about intimate matters with your parents. Parental respect, whether it was earned or demanded, didn't allow for a child to ask such personal questions. Gloria was not comfortable telling Mom that her DNA test results showed that Mark was not her father.

Gloria had power of attorney over Mom's estate. Diane, Paul, and Neil were the only children still living in the town we all grew up in. The rest of the family had moved far away by the time all of this new information about our family was coming together. Neil wouldn't have anything to do with Mom because of the pig incident and other conflicts they'd had over the years. Paul's health had been deteriorating badly, and he was having a hard enough time running his own affairs. There was no way that Gloria or anyone else was going to put the burden of taking care of Mom's affairs on Paul's shoulders. This meant that Diane and Gloria were the only two left to take care of Mom's legal and medical matters. Gloria lived about two and a half hours away from our town, but would come visit her every week to make sure things were taken care of. Diane was available to take care of the more immediate needs and participate in the regular council meetings with Gloria and the care facility's management staff.

Several months after Gloria got her test results, I asked her if she had spoken with Mom yet about who her father was. She had kept putting it off because she didn't know how to approach her. It was hard for me to raise this issue with Mom, so I understood completely what she was going through. I encouraged Gloria to talk to Mom soon; if Mom died before Gloria had a chance to ask these questions, she might lose all hope of ever knowing the truth. At that time, Mom was ninety-two years old, and we knew she could go at any time.

About three weeks later, I got a call from Gloria. I could tell by the sound of her voice that something was up. She said she had just talked to Mom and found out who her father was. They had just been chitchatting when Gloria decided it was time to do it. Out of the blue, she told Mom that she'd had her DNA tested so she could have information available for her grandchildren. The tests told her that Mark was not her father; she wanted to know who it was so that if there was a medical emergency, or if one of the grandchildren gave birth to a child with some unusual disease, they might know which family line it came through.

Gloria said that Mom didn't seem especially upset by this news. She came right out and told her that she was fathered by a man named Robert Marsh. When Gloria said she hadn't ever heard of a man by that name in our hometown, Mom said that Marsh was a soldier from Missouri who had been temporarily stationed at a military base in Chicago during World War II. She said that she and a few girlfriends had made a mad dash into Chicago to have a night of dancing and drinking, and that's when she met Marsh. Mom was married at the time with two very young daughters, and Dad was overseas, serving in Europe. Mom assured Gloria that this was just a weekend fling, not a long-term relationship.

Gloria decided to ask Mom about Paul as well, and Mom revealed that Michael Paul Keller was indeed Paul's biological father. Mark wasn't Paul's dad like she had originally told me when I first confronted her. Once again, she had not told me the truth about either Paul or Gloria when she had said they were both Mark's children. I had suspected these were lies at the time, and this revelation simply confirmed my suspicions. Now we knew that seven of the nine children in our family were not fathered by Mark Anderson.

With Mom's penchant for not telling the truth, however, we could not be 100 percent certain that Paul was not Mark's son until we could prove it through genetic testing. Some time had gone by since we had started this journey, and Paul had died recently after losing his battle with cancer, the second of the nine children to die. This presented a problem for us.

When we all returned to our hometown for Paul's funeral, Judy, Neil, and Diane had little to no idea of what had been going on with my research. Tim, Carlee, and Gloria were all on board and, except for a few pieces of information, were fully informed about what had taken place thus far.

JUDY WANTS TO KNOW

Tim, Carlee, Gloria, and I were wondering when and how we were going to tell Judy about what had been going on. Judy was the second child of Mark and Linda Anderson. With the passing of our oldest sister, Holly, Judy was now the oldest living child in our family. We were concerned about how she was going to take this. She was close to Mom, and we were sure she would not approve of the fact that we'd been gathering all of this information, much less the information itself. Judy, like Gloria, comes from that generation that disapproves of digging into personal matters, believing it shows a blatant disrespect for one's elders.

After some discussion, we decided it was time to include her in our family history mystery. We decided we'd just wait until one of us felt the time was right. A few days later, Tim and Judy were driving together, heading to pick someone up for a family activity. They had to wait for about twenty minutes for their passenger to get ready, so they started talking in the car. Tim came right out and told Judy about what we had discovered. He told her that he, Gloria, and I had all had our DNA tested, and that none of us belonged to Dad. He also told her that Gloria and I had both spoken to Mom and learned that Gloria, Diane, and Paul were not Mark's kids. He revealed that when I had originally spoken with Mom, she'd told me that Diane, Paul, Gloria, and Judy were all fathered by Mark, but later, when Gloria had talked with her, Mom had said that Paul was

Michael Paul Keller's son. And of course, we'd found out through Gloria's DNA testing that Gloria was not Mark's.

Judy listened in complete awe as Tim recounted the details of what we had discovered, saying, "Are you sure about all this? I just can't believe it!" Tim suggested she might want to seriously consider having her own DNA tested to see if Mom was telling us the truth about Judy. Tim had come prepared with a DNA test kit, just in case she consented, which she did right away. I was happy to hear later on that Judy was so open to the idea of being tested. I had been concerned she might feel it would be disrespectful to do it without Mom's knowledge.

Tim sent Judy's DNA sample to Andergene Labs in California to have Jack run the standard tests on it. By this time, Jack had become quite intrigued by how this Anderson family story was developing. Little by little, he'd learned more about our family, and he was fascinated. Each test revealed a new twist. He thought this was as good as any story you might find in a checkout-stand tabloid. The difference was that this was all truth—no aliens or Elvis Presley sightings. This was all on the up and up. Jack was more than happy to help us in any way he could.

It was no surprise to most of us when Judy's test results showed she was not Mark's daughter. Still, it was a shock to Judy, raising the same list of questions we'd had to deal with: Who was her father? What were the circumstances that led to her conception? Why hadn't Mom shared this with her? So many questions, with very few answers available to help her come to terms with this bizarre story.

At this point, we felt pretty confident that of the nine children in our family, none of us had been fathered by Mark Anderson. This only added credence to our previous thought that Dad may have been sterile. Through the indisputable proof of DNA testing, we now knew, without a doubt, that Judy, Gloria, Tim, and I were not fathered by Mark, and there was a very high probability that Neil was also not Mark's son. The other four children had an extremely high probability of not being Mark's children because of a variety of clues and information provided by Mom, which, in and of itself, was not reliable. However, by looking at photographs of the purported fathers of the remaining four children, we felt confident there was sufficient resemblance between these men and our siblings that we could safely state that none of them could claim any DNA from Mark Anderson. Two of those four, Holly and Paul, had passed away and could not be tested directly. The only thing we could do at this point was test their children to see if they carried any of Mark's DNA in their blood. The other two are still alive, and we expect that eventually we will have them tested to validate what we already believe to be the case: that they are also not Mark's children.

CARLEE WANTS SOME ANSWERS

About a year after Gloria and Judy learned that Mark Anderson was not their real father, Carlee decided to have Andergene Labs test her DNA. She was already confident that she was not Mark's daughter, but was looking for indisputable evidence. A few events happened around the time Carlee was born that raised questions as to her real paternity. Over time, those doubts began to wear at Carlee's mind until she finally decided it was time to have her DNA tested, so she could put these doubts to rest.

Carlee is the youngest of our family of nine children. Several months before she was born, my oldest sister, Holly, was dating Brian, a guy she knew from school. He looked a lot like James Dean, the famous movie star and cultural rebel. As a young kid, I thought he was so cool. He always treated me well. Even though he looked tough, he was nice to me. That was enough to make Brian a great guy in my mind.

Several months before Carlee was born, our parents decided that our sister Holly would go to Montana and spend the summer with our aunt Jean. This wasn't too unusual in our family. With nine kids in the family, most of the older ones were more than happy to spend the summer with their aunts and uncles. Most of them lived on farms, typically not more than seventy miles away from where we lived. These farms provided plenty of space for my siblings to play and work with their cousins. The aunts and uncles seemed more than happy to let them come and stay for the summer; after all, it meant more hands to help with the farmwork.

Holly's summer trip to Aunt Jean's house was a bit different, in that Aunt Jean lived a few states away. No one had ever gone that far away to stay with relatives for the summer. Another thing that made Holly's stay so different was that Aunt Jean was a single mother and had never taken in anyone's kids like the other aunts and uncles did. Nor did she have any daughters; she only had one son, who was about Holly's age. Finally, this stay extended into the school year, which had never happened with any of the other kids.

When Holly came back from staying at Aunt Jean's house, Mom had just given birth to Carlee. It wasn't long before the small-town gossip club put the pieces together and assumed that Holly had had to leave town because Brian had gotten her pregnant. Holly was furious when she heard this. She went over to see Brian and told him that if those old gossipmongers were going to talk, then she wanted to give them something to talk about. She put Carlee in her stroller, grabbed Brian, and the three of them walked up and down Main Street in front of the shops where the old gossip queens were working. Oh, Holly loved seeing the looks on their faces! Brian seemed to enjoy it just as much as Holly did.

Carlee had asked Holly several times over the years if what had been said by the old gossip club was true. She had wanted to know if Brian was really her dad. Holly had remained adamant that she was not Carlee's mom, and that our mom, Linda, was her real birth mother.

Poor Carlee carried the burden of not knowing for sure who either of her parents were for many years. Even though Carlee eventually felt confident that Holly had told her the truth about who her mother was, she continued to have doubts about the identity of her biological father. Over the years, she'd heard the three oldest girls say that a man named Peter was actually her dad. Carlee asked them to give her all the facts; they did one better, providing her with a photo of Peter, the man who was supposedly her real father. When Carlee saw the picture, she knew without a doubt that she was

Peter's daughter. She was the spitting image of Peter. But still, she lacked 100 percent indisputable proof, which could only come through DNA testing.

One day, Carlee and Mom were talking and Carlee came right out and asked Mom if what the older sisters were saying was true. Mom admitted that Peter was her father. She didn't provide Carlee with any details about how this had happened—only that he was the man who had fathered her.

With these stories in mind, Carlee finally decided it was time to find out if what she had heard all these years was really the truth. She asked me to provide her with a DNA kit that she could send to Andergene Labs. Her test was going to be different than the others we had sent; we planned to have Jack test her DNA against our mom's, to see if they shared the same mitochondrial DNA that runs along maternal lines. She wasn't planning to test to determine if Mark was her father, because she felt the resemblance between her and Peter was so close, it was enough to establish that Mark was not her father. Even so, I convinced her to have it checked against Mark's DNA profile, just so no one could ever dispute the fact that Mark wasn't Carlee's father. That way, we could focus on learning if Peter really was Carlee's biological father.

When the tests came back, it showed without a doubt that Mark Anderson was not Carlee's father, which was no surprise at all. I think the photo of Peter had cinched that one up right from the start.

The tests also proved that Linda was Carlee's mother, not our sister Holly. Although Carlee was relieved to know that Linda was her biological mom, I think she may have been a little disappointed. Carlee loved our sister Holly and was always very close to her. Holly was often the one who provided Carlee with her daily care when she was a toddler. In many ways, Carlee felt that Holly was more of a mother to her than Linda was. Nonetheless, if Carlee's biological mom had been Holly, that would have taken our family story to a whole new level of difficulty, and we definitely did not want to go there. It was already complicated enough.

So now Carlee had the scientific evidence she had been looking for: Linda was her biological mother, and Mark Anderson, the man who had raised her, was not her biological father. This allowed her to put even more stock in the idea that Peter was her biological father, although, without DNA testing, she can never say with 100 percent assurance that this is the case. But for Carlee, the circumstantial evidence is enough for her to feel comfortable with this conclusion.

WHAT ABOUT PAUL?

At this point, we knew from DNA testing—and, in Neil's case, a questionable blood match—that Judy, Gloria, Neil, myself, Tim, and Carlee were not fathered by Mark Anderson. So, that left only Holly, Diane, and Paul who had not been tested. Holly had died a few years before, and with Paul's recent death, our options for testing his DNA were extremely limited. Fortunately, we still had one option left.

Paul had one biological son, named Mike. To find out whether Paul was biologically related to Mark Anderson, we would have to go through Mike. We knew that the Y chromosome travels exclusively through the male line, so if I could get a sample of Mike's DNA, it would be a simple matter of comparing it against Mark's DNA (which we already had on file) to see if Mike's Y chromosome matched Mark Anderson's Y chromosome. The only way Mike could get a matching Y chromosome from Mark would be to get it through his father, Paul. With this one simple test at Andergene Labs, we could get our answer.

The problem was, I would have to tell Mike what had been happening—and that chances were good Mark was not really his grandpa. Mike had been very close to Mark. I worried that learning Mark was not really his biological grandfather could be a traumatic experience for Mike. In addition, some of the older sisters didn't want word of this spreading to the other grandchildren. What they seemed to be forgetting was that several of them already knew at least some of the details of our story, so I knew that word was going to get around, if it hadn't already. Nonetheless, because of their concern, I was trying to limit the knowledge to as few of the grandchildren as possible.

So, what to say to Mike? Although I'd thought of a dozen different ways to explain why I was asking for a sample of his DNA, I knew I didn't want to lie or be unethical in any way. I felt it was my duty to make sure that whatever I told Mike about my family research, it had to be 100 percent truthful. No lies, no stretching or distorting the truth. Thus, I put off asking Mike for a DNA sample for more than a year. Then one day, I saw that Mike had posted on Facebook that he had nearly escaped death on his motorcycle. After reading what happened, I was reminded that Mike is the only living male descendant for Paul's line. If we lost him, we would have no one left to test for a Y chromosome match. I knew I couldn't put this off any longer; I had to call him and tell him what I was up to.

I called Mike two days later and told him that I needed his help with a very important family project. After I'd explained what I was doing and why I needed some DNA from him, he was more than happy to cooperate. I told him the process was completely painless, and that I would cover the costs. I was quite relieved that he was so willing to help out.

Then, to my surprise, Mike told me that about a year before his dad died, Paul had told him that he had heard from someone—I'm assuming it was my oldest sister, Holly—that Mark Anderson was not really his dad. He said that it might have been a man by the name of Michael Paul Keller, and that Paul had been named after him. Although I'd already heard this story from Gloria and a few others, I'd wondered if it could have been the older sisters teasing Paul to make him think he had been adopted. But according to Mike, Paul didn't seem to think these stories were simply a case of his sisters trying to irritate him. He seemed to believe that this could be true. Paul knew at the time that he didn't have long to live, and had asked Mike to try to find out if there was any merit to the story that Michael Paul Keller really was his biological father.

Mike didn't have a clue as to how to get started, nor did he have the money to do the testing. So when I told Mike what I was doing, and that I needed his DNA for testing, he was more than happy to provide me with whatever I needed. Immediately after this call, I sent Mike all the materials he needed to provide me with samples for a couple of different DNA tests. Along with the cheek-swab kits he would need for Andergene Labs, I also sent him an Ancestry DNA kit and asked him to provide a sample of his saliva. I wasn't interested in learning his ethnic makeup, but I did want to see if there was anyone in AncestryDNA's database who might show up as a match with Mike. If I could find a close match, I could contact that person and see if they shared any close family lines with Mike. This could help corroborate our theory about Paul's biological father.

Next, I contacted Jack at Andergene Labs. I wanted to see if he would begin testing the cheek swabs that Mike was going to send him within a week or two. I asked if he could use my father's DNA profile to see if we could establish a valid Y chromosome path from Mark Anderson to my brother Paul to Paul's son Mike. If Jack could establish this path, then we would know for sure that Mark was definitely Paul's biological father. If Mike's Y chromosome did not match Mark's, then we could safely conclude that Mark was not Mike's grandfather and, by association, conclude that Mark was not Paul's father. Between the AncestryDNA test and Jack's test, I figured we would have enough proof, one way or another, to make some valid conclusions.

A few months after my call to Mike, I got the test results back from AncestryDNA and discovered that while my nephew was related to me and my siblings through our mother's shared line, he was in no way related to Mark Anderson. This meant that my brother Paul was not the biological child of Mark Anderson. Again, this wasn't a surprise; I'd expected that this would be the case. We did not find any other connections in AncestryDNA's database that would help us connect Mike to someone who might be biologically related to Michael Paul Keller.

This was disappointing, but also not surprising. AncestryDNA's database is still in the early stages of development, with millions of new names added each year. We will have to wait and see if, sooner or later, someone adds a DNA sample to the database that matches

Mike Anderson's profile. If that happens, I plan on making contact in hopes that we can connect my brother Paul to Michael Paul Keller, his presumed biological father. At least, that's the plan now; it's the only option we have left.

DIANE DISCOVERS THE TRUTH

Diane was the last child in the family to learn about the research that Tim and I were doing. We had put off telling her because we knew that if she found out Mark was not her father, she would have a very hard time coming to terms with it, and as yet, we didn't have all the answers to the questions she would surely ask us.

Mom was especially hard on the middle three children, Paul, Diane, and Neil, who were all very strong-willed personalities. Mom could never tolerate anyone pushing back on anything she said, or any decision she made, and Paul, Diane, and Neil would not tolerate her controlling behavior. Confrontations among them almost always turned into terribly explosive events, which only made the situation worse.

Where Mom could be erratic and tough in her discipline, Dad generally tended to be easygoing in nature. I say generally, because I remember there were times when Dad got tough, and there was hell to pay if someone got Dad's goat. But oftentimes when Dad got mad, Mom was behind it, telling him not to let us get away with this or that. She could get Dad all riled up, when in reality he just wanted to handle things with a little less raw emotion and force. Because of this, Diane felt a deeper sense of love and appreciation for Dad. We knew learning our family secret was not going to sit well with her.

Another reason we had decided to wait to tell Diane was because we knew she would have an almost unending stream of questions that we would need to have some answers to. The more answers I could give her, hopefully, the easier it would be for her to come to terms with all that we'd discovered. We still had a lot we needed to figure out, especially regarding her genetic line.

After I had exhausted all of my research options with the other siblings, I finally decided it was time to tell Diane what was happening. Some of the other siblings didn't want to share the information they had learned about their biological fathers with Diane; they didn't want to deal with her questions. I assured everyone that no one was obligated to share personal information if they didn't want to. While I hoped that each sibling would be willing to share what they knew about their own fathers with the others, I realized this was very personal, and those who didn't feel comfortable sharing their information certainly had a legitimate point.

With this in mind, I decided to call Diane and let her know about my own discovery; I wouldn't say anything about any of the other siblings at this point. I would offer to help Diane discover the truth about her own biological relationship to Dad, and she could do whatever she wanted with the information. I knew she would have lots of questions about the other siblings, but I decided I'd ask her to promise not to tell anyone else what I was about to tell her. I figured if I made her swear an oath of confidence, it might prevent her from calling up the others and referring to our conversation.

I started my call to Diane with some idle chitchat, then came right out and said I had something extremely personal to tell her. I made her promise not to tell anyone what I was about to share, and once she'd promised, I told her just the basics: how I'd gathered and tested DNA samples from Mom and Dad, along with my own sample; and how I'd learned I was not Mark's son. I didn't provide any more information than that.

It was very quiet on the other end of the line. In fact, the silence was getting uncomfortable. "Oh, Stevie, I'm sorry," she said finally. "I'm so sorry."

That wasn't quite what I had expected, but I realized she was feeling deeply and genuinely sorry for me. She knew how hard it had been for Neil to learn that Mark was not his father, and how many years he'd carried that burden. Now Diane was seeing the same thing happening to me.

I let Diane know that I had learned this information about five years ago, and that, for the most part, I was over it now. I shared with her that I hadn't taken it well initially—in fact, I was a bit ashamed of how poorly I'd handled the initial discovery. Diane went into nurturing mode, telling me that it was okay, that my feelings of shock and anger were only to be expected. We spent the next half hour talking about what all of this meant, and then I explained that after I'd found out, I'd flown home to confront Mom and ask her for details. Diane recalled that she'd thought it odd I came home so unexpectedly; she'd known something was up, but had no clue what it could be. Now she understood what it was all about.

I told her how I'd asked Mom about the legendary letter, and that she'd denied its existence. Diane wanted to know if Mom was willing to reveal who my biological father was. When I shared that Mom had told me the names of each child's father, there was an audible gasp from Diane. Of course, she asked me to share this information with her. We went through the list, and I provided the name of the father for each sibling, as they were given to me by Mom, including the fact that Mom identified Mark as the father of three of the nine children. Given what I'd told her about me and my brother Tim, she wasn't sure whether Mom was telling the truth when she provided the names of our fathers.

By the end of our conversation, Diane was convinced she wanted to have her DNA tested. I said I would let her know how much the test would cost, and that we'd help her find her biological father, if we discovered that it wasn't Mark. I was surprised by how well Diane took all of this news. Given her feelings for Mom, I thought she would have been a lot more upset. That's not to say she didn't get angry when she learned that Dad might not be her biological father; in fact, after she recovered from her initial shock, she began telling me about some of the terrible emotional abuse Mom had inflicted on her. With each story, her voice became louder and her demeanor noticeably more intense. I realized I needed to end this conversation, as I could hear the rage building up within her by the minute. She was making a heroic effort to control her feelings, but I could see they were getting the best of her.

I can only imagine what took place after we hung up. Her mind must have been a whirlwind of emotions, just as mine was when I first learned that Mark was not my father. Who would she talk to in order to make sense of this new revelation? Would she share it with her husband and children? Her friends? I felt badly that I had created this emotional drama for her, but I knew she would have to go through this sooner or later.

Later that day, I called my friend Jack at Andergene Labs and asked if he was up to doing another DNA test on one more of my siblings. By this point, Jack had done nearly a dozen tests for me and my family. He was fully engaged in watching our story develop. Jack told me that I was giving him so much business, I was due for a discount. "You know those places where if you buy ten sandwiches, you get the eleventh one for free?" he asked. "Well, you've sent me a lot of business with all you've done with your family, and I feel like you deserve a good discount on this test. Besides, I can't wait to see how all of this ends up." Jack may be a soft-spoken man, but he does have a great sense of humor.

I sent Diane the testing kits, and when I called to tell her what the costs would be, I learned that she had already put an envelope in the mail with a blank check in it. She told me that after our phone conversation, she'd told our brother Neil about getting her DNA tested. Neil had just rolled his eyes and asked why she was wasting her money, saying everyone knew Mark hadn't fathered any of us kids. When she heard this, she knew for sure she had to do something; it didn't matter how much the test cost. I knew then that she was in this 100 percent; there was no turning back. Everyone in the family would finally know one way or another if Mark Anderson was their father.

Jack knew there was a little more urgency to this test, so we got the results back faster than we had for any other test he'd done. I was anxious to know what Diane's DNA would reveal. I think we all knew already what the results would be, but we needed to remove any doubt.

When I'd first asked Mom who each of our fathers were, she had told me that Diane was Mark's daughter. But a few days later on a return visit to the care facility, Mom started talking about our fathers again. This second time, she told me that Diane's dad was someone other than Mark. It was late in the day and she sounded a bit groggy from her medications, so I didn't put much stock in it; I just brushed it off as one more of Mom's stories. But over the next few years, her comment kept bugging me.

When the results came back, I called Diane and asked if she was completely sure she wanted to know. At this point, she could still walk away and continue to live with her belief that Mark was her father, as she'd done for the past six decades. I just wanted to make sure she was ready to handle such a huge, life-changing event. She assured me she was ready; she had to know for sure, and it was now or never.

I told her that Jack's report indicated Mark was *not* her biological father. She was silent for a moment, taking it in. I'm sure it was just as much of a shock for her as it had been for most of us. I recalled the moment when Jack had told me Mark was not my biological father—like having the rug pulled out from under me; like everything around me was beginning to swirl at incredible speed; the nausea growing in my stomach, and the sense of rage building inside me. I couldn't help but wonder if Diane was experiencing the same thing.

The silence continued. I finally asked if she was okay. I knew it was a shock to her, but I didn't know how it would affect her. Finally, she started to talk. Her first word was "Damn!" Then she started talking about Mom. She began sharing more stories of the abuse Mom had heaped on Diane when she was younger. Each story heightened Diane's sense of injustice. She didn't come right out and say it, but I could tell what her feelings were. She was saying, "After all you've done to me, Mom, now you've done this. Even after your death, you are still taking things away from me. Now you've taken my father away from me! How could you possibly be so cruel?" Her rage continued to build as the reality of what she had just learned set in. It was personal now.

I knew there was nothing more I could say. I just needed to be there and listen as she vented her angry feelings. I needed to let

Diane voice her feelings. Finally, she said, "Stevie, I've got to go." She hung up the phone without waiting for a response, and that was it.

I knew that over the next few days she would have to come to terms with things in her own way. I'm not sure how much she shared with her family or friends. I gave her a few days to see if she would call me back to talk about things some more. When I hadn't heard from her after four days, I finally called her and asked if she needed someone to talk to.

We talked several times after that, and each time, Diane vented more about how she felt about Mom. She shared stories with me that I had never heard before. It was not an easy thing to listen to, but I knew she needed to get all of this off her chest. I remember that it had taken me about two years to finally work through my feelings toward Mom after I'd learned the truth. Like Diane, I felt Mom had taken my real father away and left me with nothing but a man I didn't even know. I understood some of the intense and painful feelings she was going through. I also knew it would be a very long and difficult road she would have to travel before she could let go of her hatred toward our mom. The abuses heaped on her were far more than anything I had ever experienced, which made me wonder if Diane was even interested in making the grueling effort to work through this, much less to forgive Mom.

During one of our conversations, I made a comment about working toward forgiveness. Diane let me know in no uncertain terms that she was not interested in going that route—at least, not yet, and possibly never. I shared my thoughts with Diane—about how we knew so little about what had happened to Mom to have broken her so badly. Diane was not interested in allowing our mom to use any excuse to hide behind, including the possibility of being abused herself, or mental illness. It quickly became clear to me that now was not the time to talk about forgiveness.

Now that we knew Mark Anderson was not Diane's biological father, the next item on our agenda became identifying the man who was. I remember very well what it felt like to not know who my father was. I struggled with this for two months before I finally flew home to confront my mother. Even then, I wasn't sure she would remember who had fathered me; after all, it had been nearly sixty years since they had had their tryst. Maybe she had been fooling around with so many men back then that she wouldn't even know for sure herself. During those difficult two months, I'd look into the mirror each morning while shaving and wonder whose blood was running through my veins. What kind of man was he? Did I go to school with his children? Did we know each other? Would I make the effort to meet this man? What would happen if he refused to meet me? So many questions. Now, Diane was going through these same feelings that I had struggled with.

When Mom finally told me who belonged to whom, she had made me promise not to tell Diane any of it until after her death. Two years had gone by after Mom's death before I finally told Diane our family secret. Because of this, there was no way for Diane to talk to Mom face-to-face and get some answers about her biological father. The only way we could try to find out his identity would be to purchase a DNA kit from the AncestryDNA website.

By this time, I was making full use of the AncestryDNA testing services. We had started by using Andergene Labs to determine who was, and was not, biologically related to Mark Anderson. We had then decided to access the AncestryDNA database to see who we might be related to. At the time, AncestryDNA had the biggest database of genealogies linked to DNA test results in the market. If you used one of their DNA tests, your information was automatically added to their genealogical database. It would then be compared to everyone else who had taken one of their DNA tests. If you matched someone from their database, it would show how you were related to them. You could see if someone matched you as a parent, a child, a sibling, a cousin, or some other relationship. Ancestry DNA had come through for me so many times before that I thought it might work one more time, with Diane.

As soon as I got my hands on a kit, I mailed it to Diane with instructions on how to use it. After the necessary waiting period for processing, Diane and I went online to see if we had any matches.

Much to our amazement, we found two, a man and a woman. Now I was nervous. This seemed too easy.

Both matches were listed as first cousins, so I knew we were on to something. (Distant cousins are often too far removed to be of any use.) I knew that both of these people might have information that would help us locate a living relative of Diane's biological father. With Diane's permission, I sent an e-mail to both matches. Within two days, the man responded, saying he didn't recognize any of the names I'd suggested to him, nor was he familiar with anyone living in the area where we had grown up. It was hard to tell if he was being honest, or if he simply didn't want to make contact with an illegitimate relative. He clearly wasn't willing to help us in any way. I decided to wait until we'd heard back from the second person before I pushed further on this first lead. If our second lead produced some results, then we wouldn't need to bother following up with this man.

The second lead didn't respond to my first e-mail. I looked at her AncestryDNA account and saw that she had not logged on to her account for more than a year. This was not a good sign. It was impossible to know when she might log on again and see our message. I also knew that when someone uses the "respond" link to send a message, the other person is notified via their iPhone that a response was sent. So, I guessed that she may have seen my e-mail and simply did not want to make contact.

I decided to see if I could locate her on Facebook. The ID she used on her AncestryDNA account was unique enough that I thought I had a pretty good chance of finding her on Facebook, assuming she even had a Facebook account. I tried it and found her within five minutes. I was getting excited again. I looked at her posting history and saw that she posted regular updates on a monthly basis. If I had to wait a month, I could handle that. I wrote a private message on her Facebook page and sat back to wait as long as it might take for her to respond. Fortunately, we didn't have to wait long. I got an e-mail reply from this second contact attempt within four days.

Our contact was a thirty-five-year-old woman named Maren. She said that she'd gotten my AncestryDNA message and had forwarded it to a cousin, to see if she thought she should respond. Maren wasn't sure if this was something she should get involved with or not. Fortunately for us, her cousin told her to go ahead and make contact. Maren called and asked me how she could help. I explained to Maren that my sister and I had discovered during the last years of our mom's life that Diane was not the daughter of the father who had raised us, and we were hoping that she might be able to help us make contact with someone in her family line.

We talked and compared notes. After about ten minutes, we were both convinced that we were close to finding answers to Diane's question of who her father was. Maren said she had a cousin named Susan who had grown up in our hometown and might know our family. With Susan's permission, Maren gave me Susan's phone number and wished me luck, adding that I could call her back if Susan didn't have the information I was looking for.

I hung up with Maren and sat back to catch my breath. I was only a phone call or two away from finding out who Diane's father was. It's interesting to be in a position where you are about to cold-call someone and tell them that you think your sister is the (previously unknown) child of an immediate relative of theirs. Of course, I wouldn't put it exactly like that. I knew that within a few minutes I would be talking to a woman who knew absolutely nothing about me or Diane. I would tell her that I was acting on behalf of my sister, and that we had some reliable proof that she is the daughter of this woman's uncle, brother, or someone close to her.

I had a million thoughts going through my head at the same time. My mouth was dry as I thought about making the call. This part of my research is never easy. The person on the other end of the line could have any number of reactions, including telling me to go away and never call her again. If that were the case, we might never know who Diane's father was. Yes, I was nervous and on edge, but I couldn't go back now. I had to make the call.

When Susan answered the phone, I calmly explained who I was and why I was calling. I could tell from her voice that she was an older woman, maybe in her mid-seventies. She very politely introduced herself to me and said that she had been waiting for my call. So far, so good. She let me talk and never interrupted me during my initial explanation, nor did she hang up on me, which was another good sign. I knew now that we had a chance with this woman. Maybe she would tell us what we needed to know.

I told her that my mother had lived in a small town in the suburbs of Chicago, and that she had had nine children. I also told Susan that before my mom died, she had revealed that the man who raised my sister was not really her father. (I didn't bother trying to explain that this was true for all nine of us; I figured this would only complicate things, and might scare her enough that she might hang up on me.)

After some discussion, I could tell that I had hit gold. This woman was going to lead me to Diane's immediate family. As we continued our conversation, I explained that my mom had said Diane's father was a man named John Davis, and that he had lived in our hometown, or possibly somewhere near Chicago. I told Susan that this was the man I was looking for; would she help me find his children, and possibly let them know ahead of time who I was? As soon as I said this, the line went quiet. I knew she was still there because I could hear her two dogs barking in the background. After about twenty seconds, I started wondering if something was wrong—that I might not find what I was looking for after all.

Finally, she spoke.

"You know what—John Davis was my father," Susan said quietly.

Oops! I was not expecting to hear that; an uncle, maybe—a cousin, even a brother, yes—but not her father. Now what should I say?

This was getting uncomfortable very quickly. I'd known that chances were good I'd end up in this position sooner or later—that someday I would be trying to explain to someone on the other end of the line that their father had had an affair with my mother. In some cases, the men had been single, so their indiscretions could more easily be attributed to the foolishness of youth. But some of them were married men, with wives and children. To learn that their father had been cheating on their mother would surely be a difficult matter to deal with. On top of that, they would also be learning that they now had a half-sister or half-brother that they had never known about. That's a lot of shocking news for anyone to hear from a stranger over the telephone.

When I'd finally caught my breath, I apologized and told Susan that I hadn't intended to break the news to her this way. She was very understanding, and seemed more concerned about my feelings than she was about learning that her father had had an affair. We compared dates and learned that Susan was already about six years old, with four older brothers and sisters, when her father had had the affair that resulted in the birth of my sister Diane. I was amazed at how kind she was, and how concerned she was for my feelings.

I had so many questions I wanted to ask about her family. Maren had given me just enough information to allow me to do some research on Susan's parents and grandparents. I was interested, but didn't want to frighten her by asking too many questions about her family members. I decided to wait and see if she would allow me to call back after she and my sister Diane had had a chance to talk and exchange information. I asked Susan if it would be okay to share her phone number with Diane. Susan was not only okay with it, she was excited! She gave me her e-mail address as well, and asked me to tell Diane to contact her anytime; Susan was more than happy to talk with her. Wow! This was more than I had ever expected.

At this point, I knew it was up to Diane and Susan to decide what kind of a relationship they would work out between them. I had done my part to find her; now it was up to Diane to follow through. When I called Diane and reported what I had learned, she was very excited. She had thought it would take weeks or longer to find anyone who might know who her father was. She never dreamed that I would find her half-sister in just about three days—and that she'd be willing to talk with her!

Later that evening, I received a blind carbon copy of an e-mail Diane had sent to Susan, saying Diane was excited to hear from Susan, and see what the two of them could learn about each other. She had included a photo of herself so Susan could see what her half-sister looked like. I hope Susan does the same. I would love to see if there are any similarities.

After everyone else had gone to sleep, I was still thinking about finding Diane's half-sister. Diane was the last of my eight siblings to learn that the man who had raised her was not her biological father. Yet, she was the first to actually make contact with her biological father's family. I had to laugh when I thought of the irony. We had been so concerned about what Diane would do when she found out about our family's secret. We knew that her feelings for Dad would make this more difficult for her than for any of the other children. Yet, hers turned out to be one of the best success stories of any of us.

This entire story about our family never ceases to amaze me. Just when I think I've got things figured out, a new twist happens, and things turn out quite differently than I expected. If I would have known six years ago what I know now, I doubt I would have believed it.

6

ADDING CLARITY TO THE PICTURE

This whole adventure of discovering our family's true identity started in late 2007. It began with a few stories and the thought that we wanted to find out who we really are, as a family. The confession by my mother that Neil was not Mark's son and the as yet unproven stories that Carlee and Holly also may not have been his daughters provided just enough information to convince me that there could very well be more to the story than any of us might have been aware of.

When we first started, the only clue we had that any of us may not have been fathered by Mark Anderson was with Neil, at the time of his accident. The differences in their blood types raised the question of Neil's paternity with my parents, which resulted in my mother admitting to my dad the real story. We feel that there are enough details that corroborate Mom's story to feel reasonably confident that he is not Mark's son. Until Neil allows us to test his DNA, we will not have indisputable evidence; only genetic testing will provide that. For Carlee and Holly, we were going only on what our mother had told us, and we knew from experience that we could never consider Mom's testimony an unquestionable fact.

By May of 2016, we had done enough DNA testing to provide us with many valuable facts and some amazing stories. The puzzle was coming together to show us a picture we had no idea existed when we'd first started investigating our family's history. Still, there were more questions we needed answers to. I wasn't quite sure how we were going to move forward with this investigation, but with each DNA test we conducted, new windows of opportunity opened up, showing us how we might be able to find those answers that had been eluding us for so long.

BROTHERS OR HALF-BROTHERS?

In 2016, Tim and I were not sure if we were full brothers or half-brothers. Mom couldn't remember if my father was Ray Jacobson or his brother, Timmy Jacobson. She was sure that Timmy Jacobson was the father of my brother Tim. In fact, that relationship had meant enough to my mom that she had named my brother after his biological father.

When I learned that I could be either Ray's or Timmy's son, my wife and I started doing research online to see if we could find some additional information about the Jacobson family. We were hoping there might be some clues to provide us with enough proof to say, with complete confidence, that I belonged to either Ray or Timmy. We were especially hoping to find photos of the brothers to see if Tim or I resembled either of the Jacobsons.

Within the first hour of our search, my wife found Timmy Jacobson's obituary. She let out a yell and laughed. We both took one look at the picture of Timmy Jacobson included with his obituary and immediately knew that Mom had been telling the truth about Tim's paternity. The resemblance between Timmy Jacobson and my brother Tim was so amazingly close, it was impossible to believe otherwise. We never could find a picture of Ray, who died as a young man, to see if I shared any similarities with him. I didn't really look much like Timmy, but with my brother Tim, there was no doubt.

I called Tim and told him what I had found. I sent him the link to the website where he could find Timmy's obituary. When Tim saw the photo of Timmy, he laughed as well. Seeing Timmy's picture triggered a memory in Tim's mind of an event that took place several years before. It hadn't made sense then, but now, with the information we had discovered, it had all become clear. Tim said that several years ago when he'd returned home to visit family and friends, he'd spent an evening at the local country fair. Our dad was there, too, along with one of his friends. Dad introduced his friend to Tim, saying, "Tim, this is Timmy Jacobson. Timmy Jacobson, this is Tim Anderson." Tim had just laughed and said, "Your name should be easy for me to remember." They chitchatted for a few minutes, and then Timmy Jacobson and Dad walked away.

Upon reflection, Tim thought this was a bit odd. Why had Dad introduced him to Timmy Jacobson? He'd never introduced any of his children to his friends before. It didn't make any sense. Did Dad know that Timmy was Tim's biological father? Did Timmy know? If Timmy did know, when did he discover this information, and who told him? With both men deceased, we will likely never know. Perhaps it was one father introducing his son to a friend, with no knowledge of the truth; perhaps it was something more.

It was bugging me that I didn't know whether Tim and I were half-brothers or full brothers. Tim and I have always been more than just brothers. From the time we were little boys, we'd always been very close friends. As kids, we did a lot of things together. In school, we had common friends, and we both played the same sports. I'm not sure why I wasn't content to know we were at least half-brothers, but I wasn't. Deep down inside, I wanted us to be full brothers, with the same mom and the same dad.

I decided that Tim and I should be tested through AncestryDNA, as we'd done with Paul's son, Mike, to determine once and for all if we were full or half-brothers. I bought two kits, one for me and one for Tim. We provided the saliva samples and sent them in. I wasn't expecting too much from this, but I'd learned that if there was a possibility of finding something I'm looking for, then it's worth taking the chance.

When I got my test results back, I'd hit the jackpot! I was amazed to find the information I was looking for. The test showed that Tim and I are listed as "Immediate Family," not cousins or any other distant relationship. The test also found a match with someone we're more distantly related to on the AncestryDNA website. It showed that person as a cousin to Tim and me, and had her listed as the granddaughter of Ray Jacobson, not Timmy Jacobson. As I looked at the relationship in the tree, I realized my mom was wrong when she'd told me that Ray Jacobson was my father. In reality, Timmy Jacobson was my biological father. I now knew that my brother Tim and I really are full brothers, not half-brothers. This was the proof I'd been looking for. Since that first discovery, several other Jacobson family members have added their DNA to the AncestryDNA database, and each of their test results have added more proof to confirm the fact that Tim and I are full brothers.

I used this same strategy with my other brothers and sisters, and discovered that my oldest three sisters, Holly, Judy, and Gloria, also share the same biological father. I originally thought that we were nine children with nine different fathers. Now the picture was taking on a whole new look and feel.

I called Tim and shared the news with him. Out of the whole family, at least we could say for sure that we really are full biological brothers. For some people, bloodlines would not make a difference; for me, it did. Although I'd told myself throughout our investigation that it didn't matter if we were half- or full brothers—as far as I was concerned, our bond would be the same, no matter what—somehow, just knowing we were full brothers helped to cement our relationship even more.

SOME NEW INSIGHTS INTO JUDY AND GLORIA'S FATHER

After Judy and Gloria discovered that Mark Anderson was not their biological father, of course they wanted to know who their biological fathers were. Mom had said Gloria's father was Robert Marsh, a military man from Missouri. Neither Gloria nor I ever believed this story; it just didn't feel right. I'd learned long ago, especially when dealing with family matters, that when something doesn't feel right, it generally deserves some more attention. I'd learned to trust my gut. When I mentioned my feelings to Gloria, she agreed. It didn't feel right to her, either.

Thinking back on the stories I'd heard, I had a hunch there might be a slight possibility that the father of Judy and Gloria could possibly be the same man who had fathered our oldest sister, Holly. I was basing this on four important facts: First, as I mentioned earlier, before Mom married Mark, she was dating two men at the same time, Mark and Sam. Both were vying for her hand in marriage. Mom had decided to marry Sam, which meant Mark was out of the picture. However, because of an argument she had with Sam, Mom decided to break off the engagement. She then told Mark she would marry him, and they wed shortly thereafter. If Mom's marriage to Mark was nothing more than the foolhardy act of a stubborn woman, proving to the man she really loved—Sam—that he didn't own her, then you have to wonder how committed she actually was to making her marriage with Mark work. As angry as she might have been with Sam, you have to consider the possibility that she'd still had very deep feelings for him.

Second, a few months after Mom and Mark were married, Mark enlisted in the army. World War II was in full swing, and if Dad hadn't enlisted, he knew he would be drafted. His thinking was that if he signed up for service in the army, he might have some say in where he would be posted, thus staying away from the front, where the odds of a soldier coming home were considerably lower.

During Mark's time in the army, he was sent to various locations throughout the United States and, eventually, over to Europe. While in the States, he was occasionally given leave to come home and spend time with his wife. But for the most part, he was gone for most of the four years he was in the military. Mom's ex-boyfriend Sam stayed behind to run the family farm while all of his brothers left to serve in the war. At that time, it was military policy to leave at least one son at home to help run family farms, so that parents would never be faced with the possibility of losing all of their sons as casualties of war. It also ensured that farms would continue to operate, providing the food necessary to feed the troops.

So, while Mark was off fighting the war, Sam was running the family farm, which just so happened to be about six miles away from where Mom was living. After four years of service in the army, Mark came home for good. By this time, three daughters had been born.

Third, a year before Mark died, my sister Gloria remembers talking with him at his home. He told her that things had almost turned out quite differently for the family. When he came home from serving in the military, something bad had happened between our parents. According to Mark, as he was returning home from active service, he and the other soldiers he was traveling with ended up stuck in some small Podunk town. He said it had something to do with processing the paperwork to get the troops released from service and back home. With tens of thousands of American troops all returning home within such a short period of time, the process of getting them released from the military quickly became bogged down. This meant Mark and his fellow soldiers were stuck in the middle of nowhere for a whole week.

To kill time, they played cards—a *lot* of cards. As I mentioned earlier, Mark was a very skilled gambler. He knew how to run a good bluff when one was needed, and ended up winning more than $4,000 in just that one week—quite a sizable sum at the time. One of the signs of a good gambler is to know when to cash in your winnings and walk away. Mark didn't exactly walk away, but he did send home the $4,000 he had won. He then continued to gamble, knowing that if he lost everything, at least he still had $4,000 waiting for him back home.

The average income of a working man at that time was about $2,400 a year. This meant that Mark had won nearly two years' worth of income that week—truly an amazing windfall for him. He planned to use the money to purchase a brand-new home for his young family when he got back home. The average cost of a home back then was $4,600, so his winnings would have practically covered the entire cost. He would have been able to do what few other young men could afford to do at the time: own a home outright, without a mortgage hanging over his head.

When Mark sent the money home, he gave our mom explicit instructions to put it right in the bank, and not spend a penny of it! He knew what Mom was like when she got ahold of money. She could do a lot of things well, but managing money wasn't one of them. My parents' lives were filled with a lot of hard times simply because Mom couldn't manage money. Now that they had $4,000, free and clear, Mom's discipline (or lack thereof) would be sorely tested.

I don't think anyone but our parents knew all the details of what happened to that $4,000. It's something that wasn't talked about in our home. The few times my siblings and I did discuss it, we did so in hushed voices, out of hearing range of Mom and Mark. I learned about this incident only because Holly told me about it during one of the times she was mad at Mark about something. But even then, she didn't provide as many details as I wanted to hear.

Some thought that Mom used the money to help her ex-boyfriend Sam invest in something. Others think that she went to Las Vegas with Sam before Mark got home and lost all the money gambling. Of course, this was all speculation; the things that happen in any marriage are unknowable to anyone but the two people in it. In all of my research, I have never found a single bit of solid evidence to explain where all that money went. Our parents were the only ones who knew the answers for sure, and they are both gone now. I doubt we will ever know the truth of what happened to Dad's jackpot.

One thing is certain, however; when Dad came home from the war, something very big happened, and Dad decided he was going to take Gloria, who was only a few months old at the time, and go away, leaving Holly and Judy with Mom.

This came as a tremendous shock to Gloria. She was sixty-nine years old when Dad told her this shocking bit of family history. She kept asking herself, Why would Dad leave Mom and abandon his two oldest daughters—just because Mom blew through his $4,000? Why would he take Gloria off somewhere far away, never to see her mother and sisters again? Gloria wondered whether Mark might have learned that Holly and Judy were not his daughters. (Little did he know at the time, but Gloria wasn't his daughter, either.) We don't know for sure whether Mark ever found out that Gloria was not his daughter, or indeed, if he ever had any idea how many of us kids he may have fathered.

The fourth and final clue comes from a poem written by my mother. I didn't discover this poem until after she had passed away. It was not written to Mark, but to another unnamed man for whom my mother had feelings. In this poem, she talks about how much she wants to be with him, and speaks of the daughter they had in secret. The child's name in the poem is Judy, the name of my second sister.

Untitled
How will little Judy feel
When she realizes the fact
That you are gone?
She won't believe that you will not be back.
She knows the love that we once shared,
How much it meant to me.
I know she'll never understand
Why I should set you free.
Poor thing, perhaps someday she'll learn
Before it's too late,
That love like ours could never be
We only tempted fate.

These four clues make for a strong argument that Sam's involvement with Mom was not over with their breakup on the dance floor. It sounds like their relationship survived long after Mom and Mark were married. By her own declaration, it became common knowledge within the family that Holly's father was not Mark, but Sam, her former fiancé.

After Holly's death, we thought we had lost any chance of proving a genetic connection between Holly, Judy, and Gloria. We had very little faith in Mom's word—that Gloria was conceived during a one-night stand with a soldier from Missouri. And with Judy preferring not to confront Mom about who her father was, we had almost nothing to go on regarding who her father was. Yet I had a feeling there had to be some way I could confirm this connection.

I decided to call my friend Jack at Andergene Labs again to discuss what options might be available to help Judy and Gloria learn who their biological fathers were. As we'd done with Paul's son, Mike, we decided to get a DNA sample from Holly's daughter, Tiffany. We had already established that Holly was Sam's daughter, so if we could get a cheek swab from Tiffany, we could compare it with the DNA samples that Jack already had on file from both Judy and Gloria. By comparing Tiffany's DNA with her aunts' DNA, we might be able to make a match. We already knew from Ancestry DNA's test results that Judy and Gloria were full sisters, sharing the same mother and father. If we could connect Holly with Judy and Gloria, we would know for sure that all three sisters shared the same father, the real love of Mom's life, Sam.

After several weeks, I got a call from Jack telling me that he had completed the tests on Tiffany's DNA sample. He had conclusive proof: Holly, Judy, and Gloria were indeed full sisters, all three, the biological daughters of Sam, not Mark.

Judy and Gloria were absolutely delighted to learn that Holly was also their full sister. Even in their seventies, they still call each other at least once a week, sometimes more, just to talk and see what's happening in each other's lives. When Holly was alive, they all found great joy in spending time together. Knowing that they are full biological sisters deepens the sense of sisterhood Judy and Gloria share, the same emotions Tim and I felt when we found out that we are full brothers.

For the past three years, we had believed we were a family with nine children, each fathered by nine different men, none of us by the one we knew as Dad. Now the story had taken on a more hopeful and happier sense of relationship. At this point, we have established that there were six different fathers rather than nine. For what it's worth, this does give us some comfort, knowing that some of us really are full brothers and sisters.

A few other options are available to us that might provide some additional facts about our family and, hopefully, remove some lingering doubts. Most of these options involve telling people who are complete strangers that we are their half-brothers or half-sisters. I'm not sure if we're ready to reach out to them just yet with this information. At this point, I plan to sit back for a while and see how things play out before taking the next step. It's time for me to catch my breath and look back in awe at where this journey has taken our family.

Section 2

Questions and Answers

7

QUESTIONS AND ANSWERS

While researching my family's story, I've shared what I've found with many of my friends. Most of them knew nothing of my family, and have never met any of my brothers or sisters. At first, I used to tell this story just to see how people would respond. I was having a lot of fun seeing the looks of shock and disbelief on their faces as they listened to the details of my story. I think it helped me realize that I was not the only one who reacted this way when I learned each new piece of information. Not only did it validate my own responses, but I also enjoyed watching my friends try to comprehend the details and make sense of this crazy story. Some of them thought I was making it up, or that it was some kind of joke. They were waiting for me to deliver the punch line, but the punch line never came. As they heard my story unfold, they kept asking for more information. They wanted clarification on some of the details because what they were hearing just didn't make sense to them.

As I continued to share this story with others, I noticed that a set of core questions came up again and again. Since most readers will likely have similar questions, I've listed the most common questions below, along with my responses.

Do you wonder if anyone thinks you've made up this entire story? It just seems too incredible to be true.

Earlier in this book, I noted that Brianne Kirkpatrick estimated that as many as 5 to 10 percent of the more than 50,000 DNA tests ordered revealed cases in which a child was not related to one or both parents.[1] That's a lot of people who are surprised by their DNA test results. I expect that over time, as tens of millions more of these home-testing DNA kits are sold each year, and the technology for interpreting those tests becomes more sophisticated, we'll see the percentage of these discoveries grow. As these cases become more prevalent, stories like mine could become fairly commonplace.

However, during the early years of my research, stories of non-parental discoveries were just beginning to show up in the public media. I have to admit that if someone had told me a story like this several years ago, I might have thought some creative liberties had been taken. In fact, I'm not sure I would have believed them at all. When you look at our family's experience, the whole story just sounds too crazy to be true. But that's the very reason I had to go after the facts. I wanted to make sure that all of the details were accurate and that nothing would be forgotten with the passage of time. This story has to be completely and totally accurate, for our posterity's sake.

I can vouch that this story is true. I've taken painstaking efforts to verify every detail, and have left a paper trail for anyone to follow if they feel they need to. As long as my siblings and I know it's true, it doesn't really matter whether others believe it or not. This book was written for the descendants of Mark and Linda Anderson. They are the primary audience. They need to know the truth about their family's real history. It would be too easy for the real facts of our story to change during the course of two or three generations into something dramatically different. I want to make sure that the facts will stand alone, and not be embellished or whitewashed with each telling. I want the truth of our family's origins to be passed down to future generations in its intact form, and not distorted in any way because of personal filters. This is the legacy of our family—who they are, and who they come from. As such, I've made every possible effort to make sure that it contains the absolute truth.

To this day, I still shake my head, wondering what in the world happened with my parents. I find myself wondering if everything my siblings and I have gone through really happened. That's why I had to go the extra mile to make sure I did all the genetic testing with legitimate, reputable companies, documenting every step of my research along the way. I needed to know that I verified, without question, as much of the information as possible. When I interviewed other members of the family to see if I was missing anything, I was careful never to lead the discussion to any forced conclusions that might not have been accurate.

In the end, I am comfortable with the conclusions I've made. Yes, it's a crazy story, no doubt about that. But it's a legitimate crazy story, and the facts back it up.

Would you ever like to connect with your blood brothers and sisters?

Most of my own brothers and sisters know who their half-siblings are. Several of our half-siblings have already died. At this time, Judy and Gloria have no desire to make contact with their half-siblings. Paul never knew for sure who his father was, and I have never been able to locate any other children of Paul's (presumed) father, so he never had the chance to make contact with any half-siblings before he died. Neil, Tim, Carlee, and I know who our half-siblings are, but we've decided not to make contact, at least for now. I'm confident that most of them do not know anything about this, or that we are related to them in any way.

Diane was able to make contact with her half-sister Susan. When I was working through AncestryDNA to locate one of her biological father's living children, I quite unexpectedly stumbled upon a living daughter. Our introduction wasn't what I had originally planned, but it turned out to be a good experience. Diane and Susan have shared some of their family information and continue to correspond by e-mail from time to time. Diane is seventy-three, and Susan is about seventy-eight. They live quite a distance from each other, so I'm not sure if they will ever have a face-to-face meeting. It's hard for many people to establish new family relationships so late in life. I think they are just fine exchanging e-mails for now.

In my case, I learned that I went to school with some of my half-siblings. Of course, I had no idea at the time that we were related. It's a strange feeling to know that I may have sat next to a half-sibling in some of my classes. I keep thinking about how glad I am that I never knew what I know now when I was in junior high; it would have been quite humiliating.

We've weighed all the pros and cons of letting our half-siblings know about our discoveries. We wonder if it's even worth creating all that drama in their lives by letting them know the truth. Some of my siblings have asked what possible good would come of it; at other times, I think they might like to know about this strange twist in their families' stories.

Since Tim and I share the same father, we've decided that if we are going to make contact with our half-siblings, we both have to agree on it. There will be no making contact if the other brother doesn't agree.

With Neil, the interest isn't even there. He's known who his half-siblings are for the past several decades. He simply doesn't care to meet his biological father's family, even though he lives fairly close to some of them. In fact, he wants nothing to do with this whole thing.

With each of my other siblings, it will be a personal decision they must make for themselves. It's quite possible that the opportunity may never come for them.

It's important to think about the effect reaching out would have on those within our biological fathers' families. The men who fathered some of my siblings were already married, with children of their own. What would the children of these men do if they found out that their father was unfaithful to their mother? Revealing this kind of personal information could be terribly devastating. On the other hand, these half-siblings might like to know we exist, and maybe even establish a relationship with us. In addition to that, there are some definite medical issues I know of that raise concerns among my brothers and sisters, and need to be addressed with preventive care and vigilant screening. I know that in my own case, the brother of my biological father died in his late forties of a heart attack. Their father died at fifty-eight; their grandfather, at fifty-one; and their great-grandfather, at forty. I've traced this line back several generations, and it appears that with few exceptions, most of the men have died younger than normal, and all but one from heart-related issues. It's good for us to know this information so that Tim and I can take preventive measures to address any potential heart problems of our own.

When I thought that Mark Anderson was my biological father, I was concerned about two very prominent medical issues within my family's health history: first was the high likelihood of diabetes, and the second was the unusually high occurrence of cancer, especially in the gastrointestinal tract. My great-grandfather died of cancer. His son, my grandfather, died of stomach cancer. Three of Grandpa's children died of cancer, two from stomach cancer. Do you see a trend here? Cancer does not run in my mother's line. Very few people on my mom's side have died of any variety of cancer, but Mark Anderson's side is loaded with it.

When I started getting a yearly physical, I mentioned this genetic tendency for cancer to my doctor. He expressed a serious concern and told me that he wanted to do a yearly test to check for any signs of cancer in hopes of treating it before it had a chance to spread. Since the most prevalent forms of cancer are stomach and colon cancer, my doctor wanted to order a colonoscopy every year, and every three years, a complete scope from both ends. These are pricey and invasive procedures that may not have been necessary after all.

I did my exams without fail for several years, until I discovered that Mark Anderson wasn't my biological father, and I was no longer tied to his family's genetic curse of possible death by cancer. One of the first things I asked my doctor was whether this discovery meant I could do away with the yearly colonoscopy. He assured me that we could cut back to once every five years instead of the yearly exam. Knowing there would be no more yearly colonoscopies for me was reason enough to celebrate. This is a prime case of the good that can come from knowing your family medical history. In our case, Tim and I were unknowingly following the wrong family tree for more than fifty-five years. It's good to know what genetic health traits you have, and equally helpful to know which ones you don't.

My eight siblings and I know that all of our biological fathers have died. The wives of these men have passed away as well (the last one died in 2015). We now know that none of the wives will be affected by learning that their husbands were unfaithful—only their children and grandchildren. At last count, I figured that collectively, with all of our biological fathers, there were more than twenty-seven children, and many more grandchildren, that could potentially be impacted by the knowledge that they have half-siblings and half-cousins because of the infidelity of their fathers or grandfathers.

So, do I think we will ever contact our half-siblings? As of now, aside from Diane, who has already made contact, I doubt that any of the rest of us will. We can get some health history from public sources like obituaries, death certificates, and the like. If we wanted to be sly about it, I'm sure we could obtain additional information through a few other ways. But all in all, I doubt that any of us will try to explain our family story to our half-siblings—at least, not in the immediate future.

What were some of the challenges you faced as you started digging into your family's story?

One challenge I had to face again and again was dealing with inconsistencies in the stories I gathered from family members. Over the years, I have interviewed a lot of people in order to record their life stories. One issue I noticed time and again was that people don't see the same event in exactly the same way. You can have five different people experience the same event at the same time, and when you ask them to recall that event, you'll find that each person provides details the others will not include, or they may describe the event completely differently. A few may even say that some of the details remembered by another person never happened. It isn't that one person is lying, or that another's memories are impaired. It's just that we all experience and remember events differently.

Police officers who gather information from eyewitnesses find that the same questions asked of different people often result in a wide variety of answers. This is because we all have mental filters through which we process what we see and experience. I often tell people that the word *history* is literally "his-story" of what was experienced. Ashley Hamer wrote an informative article for Curiosity website describing how people remember the same event differently.[2] She helps to explain the phenomenon that often makes it difficult to come up with a single family narrative.

I found this happened time and again as I interviewed my family and recorded comments from others who knew my family. I asked them to try and recall some of the events that took place in the early years, before I was born. There were some details that everyone got consistently correct, but there were others that were all over the place in terms of accuracy and consistency. It's just how the human mind works. We all see things differently.

Emotions also play a huge role in what people remember, and forget. Remember the story of how our mom sold Neil's pigs to pay for a plane ticket home from Oregon? This is a perfect example of what I'm talking about. For Neil, this was an incredibly emotional experience. He was planning on selling his pigs and using the proceeds to buy a car. To Neil, this car represented freedom and influence with his friends. Whether he realized it or not, those pigs were playing a big part in his life; they had a lot of value to him. So, when Mom had Dad sell Neil's pigs and used the money to buy her a plane ticket back home, this was a betrayal so great that Neil has never been able to forgive her for it. This experience was jam-packed with emotions for him.

But this event had little to no emotional impact on Mom. The only thing the pigs represented to her was a source of money she could use to get home. It made sense to her to sell the pigs and buy the plane ticket; it was that simple.

Another issue that often came up was how emotions can play havoc with people's memories. Highly emotional experiences sometimes "created" events that most likely never happened, especially when my mom was involved. I ran into this several times. I recall talking with one man, asking him to share some of the experiences he had with my parents. As he revealed some deeply emotional stories, I could see they were quite painful for him to talk about. His voice became louder, and he began to speak faster and use incomplete sentences. I also noticed that the details of each story began to conflict with what he had told me just five or ten minutes earlier. I could tell that some of the information he was giving me was beginning to lose credibility. I knew that I had to calm him down, or the entire interview was going to be filled with mostly useless information.

It wasn't easy for me or any of my siblings to deal with what we were learning as we continued digging deeper and deeper into our family's past. Tim and I had started this research project with the understanding that one of our brothers was most likely not Mark's son, and that there was a possibility that two of our sisters were not Mark's daughters. By the time we were done, we had discovered that Mark had not fathered any of us. This has introduced all kinds of emotional baggage that we now have to deal with.

My first reaction to learning my new status as an illegitimate son was profound humiliation. Each of my siblings has had to deal with these discoveries in their own way. I have learned that if you are interviewing someone about their life and you hit on an issue that is deeply embarrassing, the person is not going to be willing to talk about it and provide you with all the important details you are looking for. Or worse yet, they may intentionally lie to cover up the details that are such an embarrassment to them. This sense of humiliation can easily create misinformation and lead to an unreliable interview.

When I gave my older sisters an early draft of this book, I could tell from the looks on their faces and the way they guardedly gave me feedback that they were uncomfortable reading it. You have to remember that the first three children in our family grew up in a generation that listened to Connie Francis, watched *Father Knows Best* and *Bonanza* on a black-and-white TV, and made out at Lovers' Lane. The last three kids in our family, including me, grew up in a generation that protested the Vietnam War, listened to the Beatles, smoked pot, and fought the Establishment.

My older sisters grew up in an environment where you didn't talk about family secrets like this. My generation didn't have an issue talking about these things. In fact, with free love being the motto of the 1970s, I expect that a lot of children were born to fathers who were not married to their mothers. As I gathered information from the older sisters, I had to deal with their hesitation and their concern for discretion. I have little doubt that what they told me was true, but I did have to decide how I was going to deal with writing a book that included many of the details that they were uncomfortable with.

How do you think future generations of your family will react to this story?

I have put a lot of thought and consideration into this question. My biggest concern initially was what people would think of my mom. I never intended to demonize her. Yes, she had a lot of problems. I am convinced that somewhere, somehow, something happened to my mom that badly damaged her. I have some thoughts about what may have caused the damage, but I won't go into any of that here, since I have no proof. I have had no training as a therapist, and have no desire to go down that road. The fact is, Mom had some very big issues that may have had an influence on what she did. That being the case, I didn't originally feel that I ought to make this story public, but with time, my feelings on that have changed.

When I voiced this concern to my children, they were quite surprised at my hesitation. They explained to me that it's natural for this to be a big deal to me and to my siblings. We are too close to these events to be able to separate ourselves from what happened. But grandkids and great-grandkids will be far enough removed that it simply won't be a big deal. For them, it's a fascinating story and nothing more. For some, it may even become a great story to share with friends at a party. I'm not worried that this will be a big problem for Mom's family line to deal with; I expect they will handle it just fine.

More importantly, I feel that my mom's grandchildren and great-grandchildren need to know something about their genetic family lines. With DNA becoming such an important tool in dealing with genetic illnesses and diseases, it could be a fatal mistake for them not to be aware of who their real ancestors were. If they need to identify specific family diseases and end up going down a family line they think they belong to, but actually don't, it could result in a family member dying, when in fact, there is no need for that death to occur. With the remarkable advances of genetic medicine, no one can afford to not have their correct genetic information. It really is a matter of life and death.

How has this journey affected the way you and your siblings feel about your mother?

Wow! How do I even try to explain the range of feelings this experience has created? Let me start by sharing a little background information about my mom, Linda Anderson. It will help you understand how things were in our family. Linda is a fascinating study of human behavior. She died at age ninety-three. Her body didn't serve her well for the last ten years of her life, but her mind was amazingly sharp for someone who made absolutely no effort to keep herself fit or eat well during her long life.

Linda wasn't an easy person to be friends with. In fact, I don't remember that she had many friends. The few she did have were very careful around her, because they knew she could be prickly and turn on them with very little cause. She wasn't good at admitting that she might be wrong, about anything. Linda never took criticism well, constructive or otherwise. She took it as a personal assault if anyone challenged her. She seemed insecure about herself in that respect. If you made her mad, she would go into attack mode immediately. When that happened, watch out, because there would be hell to pay.

Linda also had a fascinating tendency to alter reality when it was to her advantage. In her mind, the truth was something you simply changed to fit your needs. It was that simple. She made up all kinds of stories about her past, even when there was no need to do so; it's just how she did things. Some of the stories she told were about the silliest, most insignificant things, yet she continued to create them and stand by them, no matter what. She did this until the day she died. Linda had some enormously thick filters through which she processed all of her life experiences and relationships.

When Mom was about eighty-nine, I decided I needed to record her oral history. While I was in the process of recording her memories, I remember wondering how I was going to separate fiction from fact. As a professional family and oral historian, I'm used to researching facts to evaluate the reliability of stories I hear. I knew it wouldn't be too difficult to document many of the life events Mom talked about in her oral history interview. As I began my research, I discovered that nearly everything she had told me was either greatly exaggerated or just plain wrong. She made up events with only tiny, insignificant pieces of truth as their foundation. She was remarkably good at creating an alternate reality for herself.

I have always wondered why Linda felt such a tremendous need to create this alternate reality; apparently she had done this all her life. When I spoke with her sisters, they all agreed that they never trusted anything she said. I'm not a psychiatrist, so I don't understand all of the coping mechanisms people create to help them deal with life, but I know enough to wonder if anyone really knew who the real Linda was. In fact, I wonder if she even knew. We only knew the person that Linda herself presented to us as her true self.

Linda couldn't tolerate anyone challenging her authority or questioning her integrity. This became a very big problem for her. When anyone distorts the truth with the vigor and zeal that Linda did, sooner or later, people will call them on it. I knew people who questioned her claims and sometimes even insinuated that she was lying about something. These encounters brought out the worst in her. She would say terrible things to anyone who confronted her, often physically lashing out at them. The possibility of a physical confrontation wasn't the worst; it was her ability to eviscerate someone with her words that was the most frightening thing to witness. Sometimes the verbal assault would be an outright attack on the person's character, using words and a fierce tone that would make even the most seasoned sailor cringe. At other times, she would use calmly spoken statements of fact and insinuations that left you feeling like you had just been verbally gut-punched. It was amazing to see how adept she was at doing this. She had no qualms about using whatever it took to prove herself right, no matter how wrong she might be. Simply put, Linda was *never* wrong—at least, not in her eyes!

Linda came from an affluent family. Her parents were upper-middle-class and lived a good, comfortable life. They were considered stalwart citizens of our community and were very well respected. Her father owned half of the family business that had been started by his father and his uncle. When my mother was young, and while his father and uncle were still alive, Linda's father was a salesman for the company and spent much of his time on the road, leaving her mother at home alone to care for several young children.

Linda's mother came from an affluent family, as well—even more affluent than her father's family. She grew up in the South, where tradition and social status ruled the affairs of the wealthy. Linda's grandfather was a very successful lawyer in Georgia and served in the Georgia state legislature. Linda's mother had one brother, also a successful lawyer. Her mom's parents both came from respected and affluent Southern families. Her family had hired help to take care of the children's needs when they were young. Their servants also took care of all the housework and the work around the large farm. Before the Civil War, they had had slaves. After the Civil War, the slaves were freed, but they still worked for my grandmother's family. Because of that, Grandma never needed to work and, as such, never learned how to keep house, cook, or master any of the other domestic skills that young women of that era typically learned in preparation for marriage. She was in no way prepared for the challenges of raising a family of six children, mostly by herself, and didn't know how to deal with their emotional and material needs.

From my grandmother's wedding journal, I learned that my grandparents never met before their wedding day; theirs was an arranged marriage. Reading some of Grandma's writings, I found that she was quite apprehensive about a Southern belle like her marrying this boy from the North. But once she saw him, she liked how handsome he was and felt better about the arrangement. This was how my grandparents started their marriage: two kids from affluent families who'd never met before getting married. On top of that, neither of them had any of the skills necessary to raise a family of their own.

From what I can piece together from stories shared by my mom and my aunts, I'm pretty sure their father was rarely at home. He was on the road during the week and only spent weekends at home. Because of this schedule, Grandpa didn't spend a lot of time with his own kids. On the weekends, Mom's parents were both involved with community affairs, or they would go out and spend time with friends. It's possible that the absence of her father left a mark on my mom.

All of this being said, I must say that Linda did have her good side. I honestly have many memories of good times spent with my mom. She often treated me to lunch, frequently urging me to believe in myself. She also spent hours talking with me about family members from Georgia, sharing wonderful stories about their exploits. My brothers and sisters always said that Mom liked me best. This used to irritate me no end, but eventually I came to believe that it was true. I'm not sure why she treated me so well, or perhaps, a bit better than the others. Because of this, I probably have more good memories of Mom than my siblings do. Yes, I saw her mean side more times than I care to admit, but I also saw that Mom could be tender and supportive, something most people rarely got a chance to see.

All in all, my siblings and I have mixed feelings about our mom. She was a complex person. Starting around the time of the pig incident, and up to the day of her death, my brother Neil would have nothing to do with Mom. This wasn't just because he'd found out that she had conceived him with another man. Neil always had a rocky relationship with Mom, and was often at the receiving end of her pettiness. I think learning about her relationship with his biological father simply reaffirmed his feelings of contempt for her.

Judy and Gloria haven't said much about how they feel about Mom, considering all that we've discovered over the past few years. I think what made it especially hard for them was that initially Mom couldn't, or wouldn't, tell them who their biological fathers were. We weren't sure at the time if she even knew who they were. After all, when Gloria confronted Mom about it, she was ninety-two years old. It wasn't unreasonable to think that she could have forgotten after so many years. When we finally discovered that Holly, Judy, and Gloria were full sisters, and all children of Sam, her former fiancé, we concluded that Mom must have known all along that he was their father. Regardless, it must have been terribly difficult for Judy and Gloria during the year between when Mom gave them the false names and the time they finally learned who their real biological father was.

My oldest sister, Holly, managed Mom's affairs for more than twenty years before Holly died. Holly played an important role in taking care of Mom and her affairs from a very young age. I believe this is how Holly learned so many secrets over the years, before any of the rest of us knew. Either she didn't feel they should be passed on to us, or maybe Mom swore her to secrecy until after her own passing. Whatever the case, Holly didn't reveal anything before she died.

Paul and Diane had a tough time with Mom during their teen years and much of their adult lives. They struggled with their own issues, as we all did, and never felt like they could go to Mom for comfort or any kind of parental support.

While he never had a close relationship with Mom, Paul was amazingly devoted to Mark during the last few years of our dad's life, dropping in to visit him every evening for the last three years of Mark's life, to check on him and share time together. (After twenty-five years of marriage, my parents had finally gotten a divorce and lived separately.) During this time, Paul and Mark became very close—they had a lot in common, including a shared military history—while the gap between Mom and Paul grew wider. Mark's death was hard on Paul.

When Mom told me that Paul was fathered by another man, I discussed it with Gloria. We decided that no good would come from telling Paul that Mark was not his biological father. We didn't want to do anything to ruin their last years together—although I knew Paul well enough to know that if he'd ever found out that Mark was not his biological father, he still would have considered Mark his one and only dad.

I think Mom was angry that Paul didn't stop in and spend time with her, like he did with Mark, even though they lived only a few miles apart. Before Paul passed away from cancer, he told me that he'd met with Mom to tell her he'd forgiven her for all the bad things she had done to him. I was glad to hear he was finally able to let go of his negative feelings toward her and live the last few weeks of his life without the anger and hatred.

When Diane discovered that she was not Mark's daughter, she went through many of the same emotions I went through. However, she was already carrying a lot of anger toward Mom to begin with. Mom did some terrible things to Diane throughout her life. I talked with Diane about the possibility that perhaps Mom had suffered some abuse during her childhood, or struggled with mental illness. Diane didn't want to consider these possibilities; she was convinced that Mom was simply an evil person, plain and simple.

Tim doesn't have a lot of respect for Mom. The two of them disagreed on a lot of issues over the years. When my parents got divorced, Mark refused to talk trash about Mom. Instead, he was mindful of supporting us through the divorce. Mom could see that the three youngest kids still loved Mark, and she seemed threatened by that love—as if we were capable of loving only one parent, not both. To help tip our allegiance toward her, she often spoke badly about Mark, saying he was trying to thwart her efforts to provide for us. It was painfully obvious to all of us what she was doing, and we knew what she was telling us wasn't true, but Mom wouldn't let up.

This was the wrong thing to do, especially in Tim's eyes. Tim had Mark on a pedestal. While he knew Mark wasn't perfect—not by a long shot—he also knew that our dad really cared and wanted only the best for us. So, the more Mom spoke badly about Mark, the more she drove Tim away. In the end, Tim could see what Mom was about, and learning of her infidelities only reaffirmed what he already knew. I think to him these are just more "Mom stories," that end with him rolling his eyes and saying, "Somehow, this doesn't surprise me."

By the time Mom died, she had done enough unkind things to Carlee to break any close ties they may have had. Toward the end of her life, I don't think Mom really felt a need to try and be pleasant or to make amends with anyone. She simply gave up trying. She knew that she had hurt a lot of people during her lifetime, and once said to me that it was too late to make things right. Carlee and Mom had a fairly good relationship throughout much of Carlee's life, but in her last years, Mom was in enough physical and emotional discomfort that she just quit caring and started talking mean to Carlee, which only drove her away, too. When I asked Carlee how she felt about Mom after all we've discovered in recent years, she didn't seem to have a strong opinion one way or another. "That's Mom," she said. "What can you expect?"

As you can see, these revelations about our mother's activities have affected each of us in different ways. Some simply don't care; others don't seem overly surprised. And still others feel a great sense of betrayal. How we have responded has a lot to do with each of our individual relationships with Mom.

In the end, Mom knew she had hurt a lot of people throughout her life. During one of our talks a few months before she died, Mom said, "Steve, I've done a lot of bad things in my life. You don't even know the half of it. Things aren't going to be good for me when I die. Do you still love me?"

Enough time had passed since I'd first learned the truth about my biological father that I was able to tell her that I still loved her. She was glad to hear it, saying that at least there might be one person left to speak in her defense after she was gone. I was sad for her, to realize that in her mind, her legacy was so dark.

How have these revelations affected your feelings about Mark Anderson?

The more I learn about Mark, the more I am amazed by him. Let me say right up front that he wasn't perfect. I doubt that he would ever be a serious contender for Father of the Year. He drank heavily; for more than thirty years, he was a full-fledged alcoholic. Why he and my mother ever got married is beyond me. They weren't good for each other.

I once asked Mom why she'd married Mark. She told me that she'd married him because he was dashingly handsome. Apparently, many of the other women in the area thought the same thing. She was not alone in vying for Mark's attention, taking every chance she could to be out and about, holding hands, walking side by side with Mark. She thought it quite the accomplishment to have other women seeing her with such a good-looking guy. He was what some today would consider a "trophy husband."

Mark grew up in a good farming family that took hard work and religion seriously. Although he honestly tried to enjoy farm life, he never learned to love it like his brothers did. It just wasn't in his blood to be a farmer. After he graduated from high school, he moved to a nearby city to go to college. It was his mother's big dream to have one of her sons earn a college degree, so arrangements were made for him to live with an aunt in the city so he could attend classes. That didn't work out. Mark said the aunt was old and mean, strict as a Puritan preacher; she told Mark she wouldn't put up with any shenanigans just because he was going to be a college boy. After just one day, he dropped out of college and walked more than twenty-two miles back to the farm to break the news to his mother.

After his college fiasco, he decided the military was the next best option for him. World War II was heating up, and he knew it was only a matter of time before he would be drafted. He thought he'd have a better chance of choosing what he wanted to do if he signed up with the army. He spent the next four years serving in Texas and, later, in Europe, outfitting new enlisted men.

After he came back from the war, he decided to go into sales. In fact, he was quite a talented salesman and loved the work. I remember one of his friends telling me that Mark Anderson could sell anything to anyone. He tried many times to convince me that I should try and find a job in sales, saying that if I got good enough at it, the commissions could make me a rich man. But I never had his gift; he loved people, and it showed.

Mark was a very nice person, the kind of guy everyone liked. I can't think of a single person who didn't like Mark. From what I've learned from some of his buddies, he knew how to fight and wasn't afraid to do so, but he rarely had to resort to violence. He was smart enough to talk his way out of a fight, and in the end, they'd be buying each other drinks.

Looking back on my parents' twenty-five-year marriage, I can't figure out for the life of me how they were able to stay married for so long. Maybe it was because Dad was on the road so much, away from Mom five days a week; spending just two days a week together wasn't too bad. Divorce still had a stigma attached to it back then. In a town like ours, being divorced was like wearing a big scarlet letter, bringing shame to you and your family. This would have been a heavy burden for both families to bear, especially for Mom's father, who owned one of the biggest businesses in town. I'm sure they put off the divorce as long as they could.

I am amazed that Mark never said anything about what he might have known about these matters. If he were still alive today, I doubt he'd allow me to ask him any questions. For some reason, he was quite protective of Mom. Once when I was a teenager and angry at

Mom for some reason, I said something to Mark about how impossible and unreasonable she was being. He made it clear to me that I was never to talk about my mother that way. When I think about all the things she said about Mark and the way she treated him, his reprimand doesn't make any sense to me. For some reason, he thought it was important that I treat her well and respect her as my mother.

We were never sure how much Mark knew about Mom's affairs, but it seems reasonable to think he knew something. Regardless of how much he knew, he loved us and he showed it, especially to the last three kids. I'm sure this was not an easy thing to do at times.

How do you feel about genetic testing?

This experience with my family's DNA story has helped me to see DNA testing as a remarkable tool for family history research. It has come a long way from when it was used primarily to resolve paternity suits in previous decades. As it becomes more refined and costs come down even more, and as companies continue to build up their databases of test subjects, we're going to see the science of DNA dramatically change the dynamics of the family, past, present, and future.

In my own case, genetic testing has proven that Mark Anderson was not the biological father of any of the children he raised, which was something none of us ever considered even remotely possible until just recently. It also helped us discover who our real biological fathers were. (Two siblings have yet to connect to descendants of their biological fathers.) This is huge for us. Our mother was either unable to remember or chose to hide the identities of our biological fathers. Were it not for DNA testing kits, we may never have discovered this information. Most people want to know whose genes run through their blood; it provides a sense of identity we all need and want. Matching one's DNA with someone else's helps to connect people who might never have been able to connect before.

The easy and inexpensive access to DNA testing has the very real possibility of creating discoveries that are simultaneously wonderful and very uncomfortable. Let me give you a couple of examples of what I mean.

Mark once told me that when he was in Europe during World War II, many men in his company thought they would never return home alive. It wasn't uncommon for soldiers who thought they were going to die to throw caution to the wind and indulge in all of their natural human desires. Dad said that when men from his unit (and thousands of other units like his, I'm sure) were given leave to go into town for a weekend, they got good and drunk and had sex with as many women as they could. Most of them figured it wouldn't make any difference, since they were probably going to die. Birth control was not easily available for many of these men, so there was an exceptionally high risk they'd be getting these women pregnant. It didn't matter to them. If they died, there would be no consequences. If they lived, no one back home would ever know what they'd done. They would leave whatever country they were in and return home, leaving no contact information for anyone to get ahold of them after the war. Mark also mentioned that some of the unmarried men felt badly that they might die without fathering a child. For these men, it was their intention to get a woman pregnant—in some cases, as many women as possible—in an effort to achieve some sense of immortality by leaving behind a "genetic legacy" to ensure that their genes would be passed on to future generations.

Many thousands of war babies were born as a result of this I'm-probably-going-to-die-anyway attitude. Think about it: Most of the men who fought in World War II have died by now, but their children are still alive. Then there was the Korean War, the Vietnam War, both Gulf Wars, and all of the other military conflicts we've been involved in. With the exception of World War II, many of these veterans and their spouses and children are still alive. Now that DNA testing is so inexpensive and accurate, we are beginning to see people from around the world trying to link up to their biological fathers and their American families.

It is estimated that tens of thousands of Vietnamese children were fathered by American military personnel. Life for them has generally been an unpleasant experience. Because of their mixed parentage, many opportunities to get ahead have been closed to them. Most have experienced extreme prejudice, making life very difficult. Many of them are motivated to find their American fathers and begin a new life in America.[3] If your mother told you that you were fathered by a man from a rich Western country, this would present you with some very exciting possibilities. Connecting with a man who might be able to get you out of your wretched conditions and help you come to a safer, more affluent Western country would be too good to pass up. Or maybe some of these children simply want to know who their fathers are and what their ancestry is. Either way, this has the potential of creating some very interesting, and potentially difficult, family reunions.

A second issue that should be seriously considered is what DNA testing can do for people who were adopted and have no way of finding their biological parents. For many decades, adoption cases were sealed, making it nearly impossible to find any information on a person's birth parents. The birth mother or, in some cases, the adoptive parents did not want the child to learn anything about the circumstances of their adoption.

Things have changed dramatically now. An adopted child doesn't have to bother petitioning the courts to access their adoptions records, nor do they have to resign themselves to the possibility of never locating their biological parents. They can simply buy an inexpensive DNA test from a testing company and get their DNA sample added to its database. If they closely match someone in the database, chances are good that this person is their biological parent, or a very close family member. They have a golden contact that in most cases can help them find their genetic roots. According to Brianne Kirkpatrick, founder of the Watershed DNA website, for a person trying to locate a close family member, it's no longer a matter of *if* they will find them—it's now a matter of *when* they will find them.[4]

As helpful as this is for those who have been adopted and are trying to find their biological families, this could be unpleasant news for the mother or father who put the child up for adoption.

I have a close friend who had a child out of wedlock. She knew that life would be very hard for her daughter, as she had no family support system in place to help her raise this baby. She put her baby up for adoption, adding conditions to ensure that her daughter went to a family that met specific requirements, in hopes of providing the best home environment possible. She chose a "closed adoption," which means a legal decree indicated that no one would have access to her adoption files. She felt terrible for giving up her child and was doing all she could to put this pain behind her.

My friend is familiar with DNA testing companies and what they offer. Because of this, she has refused to submit her own DNA to any of them, for fear this child might one day match up with her DNA. Reconnecting with this child (who would be in her early forties now) was not something she was ready to face. She thought it best if this child never knew who her biological mother was.

Eventually, her biological daughter did add her DNA to the AncestryDNA.com database. Even though my friend never added hers, her sister did, and out of the blue one day, her sister got an e-mail from a woman in her forties, saying she was related to her sister (my friend). Of course, the sister knew exactly what this was all about. She called my friend and told her about the phone call. My friend went into a panic and told her sister not to respond to the e-mail. While this put her sister in an uncomfortable position, she honored my friend's request.

With time, my friend reconsidered and decided that she wanted to make contact with her daughter after all. A few months later, we arranged for my friend to meet her biological daughter and her adoptive mother and siblings at our home. It turned out to be a wonderful experience. Now my friend and her biological daughter are in regular contact; even though they live more than a thousand miles apart, they visit each other and call often to exchange news.

For my friend, this has been a wonderful blessing. But this might not always be the case for others. Many women (and men) who gave up babies in their youth may not have told their current partners about it—or told their children that they have half-siblings. In the past, a person could give up a baby and be reasonably assured that there would not be any contact in the future, but this is no longer the case. For better or worse, I expect we will see thousands of children who were given up for adoption locating their biological parents and bringing a dramatic change to their family's history.

As a family historian, I've gathered many oral histories and family stories over the past four decades. When I listen to people recount stories about events that have taken place in their family's past, my attention is drawn to those accounts that don't quite add up, or just don't feel right. I've become skilled at noticing red flags. In many cases, I can tell that some of these stories were created to hide something—perhaps something considered unsavory, or a black mark on the family's reputation. This is when you might consider DNA testing and what it might provide for you. While DNA testing won't answer all of your research questions, it can certainly help when paper or digital records can't be found, or when you sense that something doesn't add up. This remarkable tool is here to stay. It's become affordable, and it's amazingly accurate. I expect that genetic testing will revolutionize family research every bit as much, if not more, than the introduction of microfilm and personal computers years ago.

If I have learned anything from this experience, it's that you need to be careful about what questions you ask. You may not like the answers that your DNA test results reveal. Are you ready to handle new revelations that might redefine the dynamics of your entire family? Have you decided how you're going to deal with the extremely sensitive discoveries that could turn family members against each other? I'm not saying that DNA testing will reveal explosive results for everyone, but if genetic secrets are out there, testing could very likely expose them. With DNA testing, it has become very easy to find what some families have spent years—and a lot of effort—trying to hide. Be ready for the possibility of surprises—and I mean big ones. Like Pandora's box, once you open yourselves to genetic testing, your world may never be the same again.

Knowing what you know now, do you think Mark Anderson was sterile?

That's the million-dollar question. Growing up, we never even gave this a thought; in fact, with nine kids in our family, sterility was the last thing anyone would accuse Mark of. It wasn't until we discovered that we had a brother and possibly two sisters who were not fathered by Mark that one of my sisters jokingly said, "Wouldn't it be funny if we found out that Dad was sterile?" By 2012, we had learned that an additional four kids had not been fathered by Mark. What used to be nothing more than a joke quietly shared among us kids had become a very serious possibility.

You have to ask yourself: How could a man be married to a woman for twenty-five years and never get her pregnant, unless he was sterile? It seems like far too many men in town were able to get Mom pregnant with just the wink of an eye, but not Mark.

I remember Mark telling me something when I was about fourteen years old. I had come down with a case of the mumps and didn't think much of it. I remember it was quite painful around my face and neck, and I missed several days of school. It seemed like a big deal to Mark; he kept after me to stay in bed and take my medicine, which puzzled me. I had been a lot sicker than this several times before and don't recall Mark ever making such a big deal about it as he did when I had the mumps. He told me that if I didn't take care of myself and follow doctor's orders, having mumps could make me sterile. *What? Seriously?* As a fourteen-year-old kid, that meant absolutely nothing to me. The last thing I was worried about was whether or not I could have kids as an adult.

But it seemed like a big deal to Mark. I asked him if he'd ever had the mumps, and he told me that he had. He'd come down with them when he was a senior in high school. By the time I had the mumps, Mark was forty-four. I wonder if he knew by then that he was sterile, and whether he attributed his sterility to having mumps as a young man. (By the way, it's quite rare for a case of mumps to cause sterility, so I'm sure if he was sterile, it's doubtful that it was caused by contracting mumps when he was well through puberty.)

Given what we know now, if anyone asked me today whether Mark was sterile, I would have to say a definite yes. How could I think otherwise? Of course, we will never know for sure, as he was never tested. Men from his generation didn't like to entertain such questions. It was an affront to a man's masculinity to even consider being tested for that kind of issue. But I can't help but wonder if Mark ever considered the possibility.

Do you think your parents had an agreement to have other consensual relationships outside of their own marriage?

I've had several people suggest that maybe Mark knew he was sterile, and because of this, they had an agreement between the two of them that Mom could get pregnant with another man to give Mom the kids she wanted. But Mom was never one of those women who longed to have kids, much less nine of them. Each of us just came and she dealt with it in her own way. I seriously doubt that her adventures with other men represented an effort to address some kind of longing for children that she might have had, to make up for Mark not being able to get her pregnant.

Another reason this idea of an agreement doesn't make sense to me is because my oldest sister, Holly, was born a little less than a year after they were married. I seriously doubt Mark would have had any thought at this point in his life that he might not be able to father children. The next two children came in quick succession, within about fourteen months of each other. There aren't a lot of years between any of us children. I just don't think there was ever a long-enough stretch of time between any of us that might have led Dad to wonder if there was a problem with his fertility. As far as he was concerned, all was well, and he had a big family to attest to his manliness. It also seems unlikely that Dad would have allowed Sam, the man who was competing with him for Mom's hand in marriage, to be the guy to get her pregnant. Even more improbable was the idea of so many other men taking part in this arrangement. Enlisting the help of all these men simply seems too far-fetched to be a serious consideration.

Some have suggested that maybe Mom and Mark were swingers—couples who traded spouses for an evening of fun and entertainment. I suppose it's possible, but it doesn't seem likely. I seriously doubt there were any agreements between my mom, Mark, and any other man (or men) to help Mom have the children that Mark could not provide her.

Why did you decide to go public with this very private and personal story?

There are several reasons I have decided to share this story with others.

First, this is family history in the most fascinating sense of the term. Although I may not have wanted to share it initially, as a family historian, hiding this story seems wrong. If there is one thing I've learned over the past forty-plus years of researching families and recording oral histories, it's that literally every family has their secrets. Some of them are so terrible that it's probably best to keep them hidden until they finally fade away into oblivion; there is no good that can come of preserving them. But I would suggest that comparatively speaking, there are very few stories that fall into that category. Nearly all stories of family tragedy have something of value to offer; it may take a few generations of separation before those stories can come out without causing too much trauma, but with time, even some very ugly stories can have some redeeming value.

Our family story is no different. I feel like my family is like a big pot of minestrone soup. We have a wonderful mixture of veggies, seasonings, pasta, stock, and whatever else you want to throw in, with incredibly satisfying results. My wife often tells me how surprised she is that, for better or worse, we have so much diversity and history in my family, and yet we still love to get together and enjoy being with each other. Sure, we've had our rough times, but we still appreciate our family bonds, and we really do love each other dearly.

At first, I was leery of passing this story on to future generations. Is this the legacy I wanted to leave for them? But as I was talking with one of my sons about it, he told me that it was a bigger deal to me and my siblings than it was for any of the grandchildren. To them, it was an intriguing story and nothing more. I realized that our story has a lot of fascinating information to offer future generations, as well as an entirely new genetic history—definitely worth putting on paper. It's not your typical legacy, but it's a legacy nonetheless, with some good lessons to offer future generations.

The second reason I decided to share this story came about through keeping a journal, which I found very therapeutic. When I first learned that I was fathered by another man I never knew, it left me speechless. Then, learning that all of my brothers and sisters were also fathered by someone other than the man we all called Dad—it rocked me to the very core of my being. This was a huge deal for me. For a while, I felt like I was someone's mistake. That kind of thinking can have a devastating effect on anyone's self-esteem. What happened was no fault of my own. I had absolutely no say in how I or any of my siblings became part of our family. It was 100 percent out of our hands.

I once asked Tim what he thought about all this stuff we've discovered. He shrugged and said that it wasn't a big deal; he wasn't surprised by what we'd discovered. He knew Mom well enough that none of this was really all that shocking to him. I was surprised by how unaffected he was by all of it. I tried to handle it like Tim did, but found that I couldn't do it.

The way I dealt with the revelations was to write. I wrote page after page of journal entries with each new discovery we made. I wrote almost daily, thinking that my writing might help me come up with answers that would create some sense out of what I was learning. By the time I finally came to terms with things, I had filled up a lot of journal pages. I knew that if I could write through all of this, I would find something to provide a sense that things were okay.

After reading through all of the journal entries I wrote, I realized that I had quite a story. I've read a lot of accounts detailing how people have used DNA testing to discover that they were adopted, or to find out who their real parents were, but I've never read about anyone using DNA to discover a family secret quite like ours. I thought it was compelling enough that others might find some value in reading about this experience. If nothing else, they could take some comfort in the fact that many families have secrets they only discover through DNA testing.

By the time I'd worked things out and resolved my angry feelings, I found that this story was kind of funny. In addition to writing my way through to a resolution, I also shared my secrets by talking to a trusted group of friends. With each telling, I found the reactions were all pretty much the same. At first, listeners were shocked, and then amused. For some reason, letting this go public to a small group of people gave me a great sense of liberation. From there, I expanded the limits of who I shared this with. I began speaking at family history conferences and teaching classes about using DNA to unearth valuable family stories. I soon realized that no one I was talking to saw me as some pitiful bastard. Sharing this story with the general public wasn't as threatening as I'd originally thought it would be. After a while, I decided I would write a book about it and share our story with anyone who might be interested in learning how DNA can play a big role in discovering possible secrets within their own families.

Third, I had to resolve the feelings I had about my mother. I didn't have angry feelings toward Mark; I felt like he was the victim, in the sense that most of this appears to have been done behind his back. I have little doubt there is a lot more to this that I don't know, but regardless of my feelings toward Mark, I ended up focusing all my angry feelings toward Mom.

Despite Mom's socially unacceptable behavior, there was still a strong bond between us. My childhood was quite difficult; as a young boy, I didn't have much self-confidence and struggled with several other issues. Mom made it a point to spend time with me and continually remind me that I could do anything I wanted to do. She told me repeatedly that there were no limits to what I could do. I'm not sure why she did this, and I wonder if she really believed the things she was telling me. Regardless, she continued to encourage me to aim for the stars and believe in myself. It seemed so out of character for her, but her words clicked, and I believed them. I trusted what she was telling me and took her words to heart.

When I was sixteen years old, she moved me and my younger brother and sister from our home to another state. I was not at all happy with this move at the time, but I eventually realized that this decision had a remarkable impact on my life and the lives of Tim and Carlee. No matter how mad she made me or how selfish her motives might have been, I always felt a sense of deep gratitude to her for believing in me and for moving us to another state. Countless opportunities were made available to Tim, Carlee, and me with this move. We enjoyed a great school with enormous resources, exposure to many new cultures, job opportunities that did not exist in our former small town, and friends who were not interested in doing drugs, to name just a few benefits. This move changed our lives in more ways than we will ever know.

After struggling with the intense anger I felt toward Mom after the revelations, I decided I had two options: I could get years of therapy; or I could face the issue head on and come to terms with what Mom had done. Whatever avenue I chose to follow, I knew I needed to deal with these feelings. I refused to go through life letting this anger eat me up inside, turning me into a bitter old man. I knew I wouldn't like the person I would become if I didn't resolve this issue.

With time, and a lot of writing, I was finally able to figure things out and let go of those feelings. My research has helped me to understand my mom a lot better. My sister gave me a collection of poems written by my mom at the time all of this was going on. It was strange to read them, realizing that most of what she wrote about didn't involve Mark, but other men she felt close to. Reading her poems gave me insight into her mind and heart.

The fourth reason I decided to share our story is because I believe it gives a great deal of legitimacy to the power and value of DNA testing, and what it can do for family history. Using DNA to research one's family history can do a lot more than just reveal the ethnicity of one's ancestors. A personal account like this one might help others see the value and range of what DNA testing can do.

If you could change any part of this experience, what would you have done differently?

I think there is only one thing I wish I would have done differently. I wish I would have started investigating things a lot earlier. Knowing what I know now, I would have taken more chances and asked more questions, even the ones that could have upset Mark. I probably wouldn't have stopped until he got right in my face and told me it couldn't go on any longer. I would have liked to have asked Mark to tell me his version of what went on and why he did some of the things he did. I would like to know how much he really knew.

I think I might even have had the courage to approach my biological father and talk to him while he was still alive. Of course, I would have promised that neither his wife nor his children would know about me and what he said, at least until after he and his wife were gone. Those are the answers you cannot learn from DNA samples. I have so many questions that I would love to ask those who would know the answers. I'm not sure they would have been willing to speak with me, but at least I could have tried.

Something else I would have done differently is that I would have started gathering DNA samples from all of my siblings and a few other family members while they were still alive. At the time of this writing, two of my siblings and both of our parents have died. I described our last-ditch effort to get a DNA sample from Mark at the funeral home. If you don't want to take the very real risk of missing out on getting a good DNA sample from a family member, get the samples before this point; it's so much easier when they are still living.

With my two deceased siblings, I've had to work through their children to get DNA samples. This was a lot more complicated, and definitely more expensive, because of additional tests that needed to be done. If possible, start considering what you need to do to gather and preserve DNA samples from your living family members. You might also start thinking about what questions you want answered through the testing of your DNA samples. Then look around and see what companies offer the services you are looking for, and how much they charge.

Finally, I think I would share information with my sister Diane sooner than we did. We waited as long as we did for two reasons. First, Mom made me promise that I wouldn't tell Diane until after Mom died. Second, I wanted to make sure that when we broke the news to Diane, I would be able to provide her with as much information as I could reasonably find, and that the information was correct. None of us were sure how well she would handle news like this, so I wanted to make sure she had all the facts, so she could take it in and come to terms with things. I realize now that we probably waited too long. The longer you wait to share information about these kinds of DNA discoveries, the harder it is for someone to come to terms with what you're telling them. To be perfectly honest, I'm not sure that there ever is a "good time" to break news like this to someone. It will likely be a shock regardless of when you tell them.

I don't think there is anything else I would have done differently. For the most part, I'm pleased with the course we've taken and how things have progressed.

Why did you wait until you were in your late fifties to start doing this DNA testing on your family?

The biggest reason we waited so long was because both of our parents were still alive. For years, Tim and I had talked about how useful it would be if we could get DNA samples from our parents to create a standard against which any of us could be compared to see if Mark really was our biological father. Even though we had talked about it, neither of us wanted to approach our parents to ask for DNA samples. If they had had any clue why we wanted the samples, we would never have gotten them. This was just too touchy for anyone to bring up. No one had the nerve to let them know we seriously wanted this information. It wasn't until Mark's death that Tim and I realized if we didn't do something right away, we might never get a second chance. We no longer had the luxury of putting this off. As it was, we waited more than two years after collecting the DNA sample before we made any serious effort to begin testing.

A second reason we waited so long was that ten years ago, DNA testing was a lot more expensive and a lot less sophisticated than it is now. By the time we started testing our family's DNA, the companies we used were providing information that fit our needs exactly. Some of those services were simply not available even five years before. Keep in mind that when we started our adventure into researching our family's unique history, we had no idea that my parents' activities included so many other people and their families. Had we known how far this story really went, Tim and I might have moved on this a lot sooner.

Once we started learning for certain that there were several siblings who might not have been fathered by Mark Anderson, it opened a whole new world of possibilities, and that included Tim and me. Suddenly, this story took on a new sense of urgency, and we could no longer tell ourselves that it could wait. We needed to know some answers, and we needed them now, before any other family members died.

If we had started earlier than we did, Dad would have been alive, and I could have asked him some very specific questions. Although I doubt he would have been willing to talk to me about it, it would have been worth a try. All the DNA tests in the world will never tell us why our parents did what they did.

Tim's and my biological father died less than ten years ago. If we had started our research earlier, we might have had the opportunity to meet with him and ask him some questions, but this will never happen.

It's easy to say now what we would have done differently if we knew then what we know now, but you don't get the opportunity to go back and do things over. If we could, things might have turned out very differently.

If you could see your mom and Mark again and talk to them for one hour, what would you ask them?

Oh, where do I begin? One hour would hardly be enough.

I guess my first two questions would be, "What the hell were you thinking?" and "Why did you do this?" I would ask these two questions to both Mom and Mark. If it was simply a matter of Mom being unfaithful to Mark, then all of this would make a little more sense to me. But if Mark knew what was going on with Mom and the other men, and was complicit, then that would throw all logic to the wind as far as I'm concerned. I would follow up by asking Mark how much he knew about this, and when he was made aware of what was happening. I would like to see if he could give me some clarification as to how much he was a part of this, and what role he might have played. I would love to know what set all of these events in motion.

Next, I'd like to know how much, if anything, the spouses of these other men knew about their husbands getting Mom pregnant. If they knew, and were silent about it, much less gave tacit approval, then that brings this story up to a whole new level of craziness. It's hard to believe that none of them knew; if that was the case, then I'd like to ask how they were able to keep all of these secrets.

My next question would be whether Mark ever knew that any of us were not his children. It's obvious now that he knew the truth about some of them. Did he know the whole story? Did he know that none of us were his? I would also like to ask Mark whether he ever considered the possibility that he was sterile. If so, did he ever go to a doctor to see if he was?

Finally, I'd like to know if all the men involved knew about us. In other words, did Mom tell these men that she was pregnant by them? If they knew, what did they do? What were their reactions? Did they run away from taking any responsibility, or were any of them willing to do something that would show some sense of concern and compassion for Mom and their unborn child? Answering these kinds of questions would add depth to this story and offer a glimpse into the personalities of those involved.

How has all of this affected your own sense of self-esteem?

I'm fine with my newfound identity now, but I have to admit, it took a few years for me to get to this point. When I first found out that the man who raised me was not my biological father, I had some very serious questions about who I really was. Keep in mind, for the first fifty-eight years of my life, I had no clue that Mark was not my biological father. I was totally unprepared to deal with the truth when I finally discovered it. I remember telling Jack Anderson from Andergene Labs that if he found out I was not Mark's son, I would be okay with it; I'd be able to handle it just fine. Boy, was I wrong! What was I thinking?

When I finally discussed this matter with my mom, when she was ninety, I wasn't sure she would even remember who my biological father was. Although she was amazingly sharp for her age, she might not easily remember something that had taken place more than half a century before. For all I knew, this might have been something she didn't want to remember. If she had been hanging out with dozens of men at the time, there might be no way to identify him. At that point, I thought I might never find out who my biological father was, much less what he was like. This was hard to take and compounded the mental trauma.

The sheer flood of questions that filled my mind was quite literally overwhelming. Most people have a pretty good idea who they are and what their family's history is, at least regarding their parents and siblings. In a single moment, everything I believed to be true about myself vanished, and I had no idea what I could believe anymore. I looked at myself in the mirror and no longer knew who I was. I had always been an Anderson. I grew up knowing my aunts and uncles and all my Anderson cousins. I loved them all and had remarkably strong family connections with them. I had spent more than forty years researching my Anderson family line back into the early 1600s, in Norway. This history was deeply ingrained within the very marrow of my bones. My children and I were intensely proud of our Norwegian ancestry.

When Jack gave me the results of my DNA test, all of that was taken away from me in an instant. I no longer belonged to my father's family. Even worse, I had no idea *who* I belonged to. My son's initial comment was right: I really was a bastard. I was going through a very real identity crisis. Experiencing this gave me a good idea of what someone must feel like when they learn they were adopted. The feelings of shock, betrayal, and self-doubt are a lot more real and traumatizing than I ever thought they would be.

I was also so angry at my mother. In all my years, I don't remember ever being so angry at anyone. In my mind, Mom was no longer the mother who took time to teach me to believe in myself. Instead, she was someone I wanted no part of. I'm a bit ashamed by how judgmental I was when I first discovered her secret, but those were my gut reactions, and they were very real and very painful. Like I said earlier, I did not handle it well at all. Looking back at that day, I understand that I simply did not know enough about Mom and Mark's early marriage and what was going on in their lives to be able to make a rational and fair judgment of the situation. During those first few weeks, I didn't even want to talk to Mom; I was afraid of what I might say to her.

During the last several years of her life, I'd been in the habit of calling Mom every Sunday evening to see how she was doing and share news of what was happening with the family. I lived about 1,400 miles away from where she lived, so I didn't get to see her very often. For the first few months after I found out about her secret, I struggled with those calls. It was hard to pretend that I was enjoying our conversations. I knew what she had done, and that she had no idea I knew the truth. I'm sure she realized something was off; she always knew when something bad was going on in my life.

I worried that she might ask me what was happening, wondered if I would have the courage to tell her what I had learned and how I felt about her. I hated the person I was becoming. I hated having these feelings toward my mother.

On the other hand, my love and appreciation for Mark dramatically increased with each new fact I learned. I don't kid myself: I know that he must have played some part in this, but I have very few clues to give me a good assessment of how much he knew and what his involvement was. From what facts I've been able to gather, it appears that he was betrayed—at least, with the first three children. After that, his knowledge and potential involvement is anyone's guess.

Over the past few years, we have confirmed the identity of all but two of our biological fathers. (Although we don't have definitive proof for the last two, we are confident we have identified the correct men.) My siblings have told me that they all feel Mark Anderson is, and always will be, our father, regardless of who our biological fathers might be. I agree completely. Mark Anderson will always be my father. He is the man who raised us and loved us, whether or not he knew he wasn't our biological father. I respect him for that, and for so many other things.

I am fine with who I am now. I'd be kidding myself if I said that I haven't changed, or that I feel the same way about myself as I did before I learned this secret. In my mind, I'm not any different with this new identity. But biologically and, to some degree, emotionally, I know that I'm not an Anderson, and that still bothers me a bit, although not as much as it used to.

I would like to know more about my genetic and medical history. For the sake of my own health and the health of my children and future posterity, it would be helpful to have a genetic profile of my Jacobson family line, which will be possible with the advances that are being made in the field of DNA and genetic engineering. For that reason, I'd like to make a connection with Timmy Jacobson's family someday, but not now. The time isn't right for that.

Are you or any of your siblings interested in learning more about your biological fathers (e.g., their personal traits, obtaining photographs, etc.)?

Some are and some are not. I don't think Neil is at all interested in learning any more than he already knows. Since Neil grew up sharing a part of his life with his biological father, he already knows a lot about him. Most of the rest of us didn't have an opportunity to work with, or spend any time with, our biological fathers, so we have fewer memories and less knowledge about them.

Holly knew her biological father's family enough that she didn't really seem to have a need to get to know them any more than she already did. Gloria and Judy don't seem interested in learning anything more about their new family (they share their biological father, Sam, with Holly). They are fine leaving well enough alone.

Carlee knows her father was a man named Peter, but she's never met him, and knows almost nothing about him. We have never been able to locate him or any of his current family members. When he lived in our town and met our mom, he was single and didn't have a family. According to our older sisters, who knew him very well, Peter was from Germany and worked in our grandpa's factory alongside Mark Anderson. After he was let go, he seems to have disappeared. From what I can tell from my research, no one knows where he went. He moved away from his uncle's home and simply left no word as to where he was going.

I recently spoke with someone who knew Peter. She said that just before my brother Paul passed away, Peter came to visit with him. Apparently, one of Peter's close family members in a neighboring town had died, and Peter had come back for the funeral. Peter stopped by to visit Paul and wanted to know how each of my older sisters were doing. He apparently remembered them very well and was interested in what had happened to them since he'd last seen them so many years ago. This person didn't know if Peter asked about the daughter he'd had with our mom, or if he even knew that he had fathered a daughter. This took place about six years ago, so we don't know if Peter is still alive or not.

Carlee has done some DNA testing with AncestryDNA to see if her DNA matches with anyone in their database. She was hoping that if Peter had any children or grandchildren living in the States, perhaps one or more of them might have had their DNA tested through AncestryDNA and they might be able to make contact. So far, no one has come up as a match with Carlee. She still hopes that someday one of Peter's children might be added to AncestryDNA's database and they can find each other. Carlee is definitely interested in learning more about her biological father and his family.

As for me, I would like to meet someone from Timmy Jacobson's family someday, although I'm not sure when I'll be ready, or how well this information will be received. I would like to see photos of the man who fathered both me and my younger brother, Tim. We only have the one photo from his obituary now. Although he's eighty-two in this picture, he still looks very much like Tim. We'd like to see photos of him as a young boy, a teenager, and as a young adult.

I would also like to learn about his grandparents and aunts and uncles. I've done extensive genealogical research on Timmy's family line, and have learned a lot about our shared ancestors. I know how most of them died, where they went to school, what they did for a living, and some other interesting things, but I'd like to learn more about what Timmy and his siblings and parents were like.

I found a man who worked with Timmy ten or fifteen years before he died and he shared a few stories about him. I'm sure his kids could tell me so much more if they were willing to talk with me. I don't know whether Timmy wrote down any of his own memories, but if he did, I would give almost anything to read what he wrote. I just want to learn more about this man who is a part of me.

In addition to the stories, Tim and I would love to know something about his family medical history. Although I've learned how Timmy, his parents, and his grandparents died, I'd also like to know additional details about their health not recorded on death certificates.

With all the family histories I've gathered over the years, I have learned that people need to have a history of their own to connect to.

They need to know whose blood it is that flows through their veins, and learn the stories of their people. We need to know who we are and what we're made of. This is what makes me want to eventually make a connection and learn more about my biological father's family.

If you could legally prove your relationship to another man, you might have a right to receive an inheritance. Would you ever consider doing something like that?

No, never! None of us are in the least bit interested in doing anything like that, nor was that ever our intention when we started this investigation. Regardless of whether any of us have a legal right to an inheritance, we would never believe we have any right to move in on anyone else's inheritance.

You have proven through DNA testing that the man who raised you is not your father. Would you be interested in getting the children of Timmy Jacobson to donate DNA samples so you can remove any doubts that their father is, in fact, your biological father?

I can only speak for myself, but I feel that we have already proven through AncestryDNA's testing services that Tim and I come from Timmy Jacobson, and not his brother, Ray. A few years after doing the testing with Andergene Labs, I sent in a saliva sample to AncestryDNA and learned that my DNA was connected with some of Timmy Jacobson's family members.

If the possibility presented itself, I would certainly be willing to collect DNA samples from our half-siblings, in case some new technology becomes available that could possibly tell us more about our family lines. Of course, the problem is obtaining samples without compromising ethics, and without telling them everything I know. When we got samples from our mother, I didn't feel bad about not telling her that we wanted her DNA to find out if Mark was our real father. We had a right to know the truth about ourselves, and knew it would never come from her if she knew what we were up to. She was the only source still living who could provide us with any real information.

But with my half-siblings, I cannot ethically justify tricking them into giving me a DNA sample. If I ever get samples from any of them, it will be because they provided them willingly, with knowledge of what I'd be using the samples for. I would be willing to tell them the part of this story that deals with them and their father, but not the details about my siblings.

So, yes, it might be valuable to take that extra step and find out through yet another avenue that our conclusions are correct. If the cost of DNA testing continues to go down, and I eventually feel comfortable approaching fellow descendants of my biological father, then I would definitely be interested in double-checking our conclusions.

Will you ever make this information known to the descendants of your biological father's grandchildren after your half-siblings have all died?

At this point in time, I would prefer to leave it to my children to share this information with Timmy Jacobson's grandchildren. With each generation of separation, the sting of what happened will be significantly reduced. I don't think it would be a problem for Timmy Jacobson's grandchildren to learn this secret. I don't think I want my children to tell them the whole story about my siblings, however. These grandkids could very easily still be in our hometown where the grandkids of my siblings' biological fathers also live. I prefer to keep it within reasonable limitations and respect the privacy of the living where at all possible.

What did your siblings think when they first read this book?

I made it a point to have some of them read an early draft because I wanted to make sure I was correctly representing their responses to what we learned through this investigation. It was interesting to talk with them after they read it. Each of them had an emotional response. I think Carlee expressed what many of them were feeling when she said, "I finally finished reading your book. I didn't like it!"

I was taken aback by her response, so I called her to find out more. At first, I thought she didn't like it because I had the story all wrong, or maybe she thought I was blowing things out of proportion, or being too harsh on Mom. After some discussion, she finally came right out and said she didn't like it because it brought back too many unpleasant feelings. She told me that during my investigation, when I called her with each new discovery, she was fascinated to discuss each piece of the puzzle, figuring out how it all fit together. We would talk about each revelation and see what memories we might have to help explain why that event took place.

But once all the pieces were put together into one big narrative, Carlee said it was simply too overwhelming. It all sounded so terrible. It brought back memories she had forgotten and didn't want to be reminded of again. I think some of the other siblings who read early drafts agreed.

I know I also felt discomfort as I reviewed each draft. With each new discovery I described, I felt a sense of sadness that none of us were able to grow up in a normal family, with each child conceived by the same mother and father. I kept thinking how grateful I was that I hadn't known about any of this until much later in life, when I was better equipped to deal with the information I was discovering. As a man in my late fifties, I had long since given up the notion of our family being anything like the Cleavers, but I still loved my siblings and thought that we'd all turned out pretty well in the end. Nonetheless, it's been a life-changing shock to discover the truth about our family. Reading about it hasn't been easy for any of my siblings. I expect that the next generation of our family will not be as deeply affected by this story. Time and space will separate them from the impact of what we have learned and personally experienced. Maybe that is what people mean when they say that time heals all wounds.

Of course, people outside the family who read this book experience none of these feelings. To them, it's nothing more than a fascinating family history. Many are surprised to see how successful we have become, given the unusual origins of our family structure.

How have these revelations affected the relationship you and your siblings have with each other today?

Even though we have all faced some significant challenges throughout our lives, as most families do, we have always been a relatively close-knit family, with the possible exception of Neil. I think discovering at such a young age that Mark was most likely not his biological father has had a big impact on him. For most of his life, he probably experienced emotions similar to what an adopted child might feel. I wonder whether he felt that he wasn't like the rest of us. The fact that Neil and Mom never got along very well created an environment woefully lacking in any sense of love and nurturing that a child needs growing up. I wouldn't be surprised if Neil never felt he was a part of our family. He comes to most reunions and some special events, but he doesn't stay long. I'm not sure when, or if, that will ever change for him.

The rest of us take great pleasure in visiting each other, and make it a point to get together as often as we can. Our little town outside of Chicago is still our home, and we try to meet there when we can. Some of us keep in close contact by calling weekly, sharing news and enjoying time spent talking with each other. Several people have told me that given what we've gone through as a family over the past seventy years, they find it amazing that we enjoy being together so much. I agree; it is amazing, and I love it.

How much did it cost to do all of this DNA research?

The cost of finding the real story of our family is still undetermined. As a group, we've paid a total of about $3,500 thus far to learn the information we have now. Additional costs will vary depending on several factors.

The decision to invest in finding the truth about our family was of the highest importance to me. Because of that, we decided to work with Andergene Labs in Oceanside, California, a well-established genetic research company with a top-rated reputation, to determine who was, and was not, genetically related to Mark Anderson. Once we determined that Mark was not our biological father, we went with AncestryDNA to see if we could determine who we might be related to. We chose AncestryDNA because they had the biggest database, thus giving us the greatest possibility of finding a match. Between these two companies, we were able to establish that Mark Anderson had not fathered any of the nine children, and were able to connect all but two of our siblings to their biological family lines.

Do you plan on doing additional DNA testing to search for more family history information?

We will probably not do any additional testing. What we've done so far has told us what we need to know at this point. This could change if we make contact with any of our half-siblings, or if we choose to start searching for medical information in the genetic lines of our biological fathers. Those two factors could take us in a whole new direction, and would most likely add a significant amount of expense to our search. We really don't know what kind of advancements will be made in the near future. If the last ten years is any indication of the pace of change, then I can hardly wait to see what the next ten years will bring.

No scientific test results can tell me why people did what they did, but if DNA testing gets to the point where it can provide me with additional insights and information, then I would definitely be interested in pursuing it. (For example, it would be fascinating to learn whether there was something in my mother's genes that led her to be so sexually active.)

After having gone through all of this, what advice would you give your readers?

There are several pieces of advice I would give to anyone who reads this story. First and foremost, I would say that we all need to be careful about how we judge people, especially in our own family.

The entire time I was writing this book, I wondered whether readers would think my mother had no morals. When I first discovered that she'd had an affair with another man that led to my birth, I was shocked, and angrier than I'd ever been in my life. Eventually learning that she had conceived all nine of her children with men other than Mark was overwhelming. But after a long time, my emotions cooled down and I began seeing my mom in a relatively rational and not quite so emotional way. It's easy to make quick judgments about a person when you have only some of the facts.

I guess the point I'm trying to make is that our home environment and personal experiences shape who we are, often having a much bigger impact on our lives than we realize. My mom didn't share a lot of stories about her home life, and never shared details about her personal struggles. She was never one to admit faults or talk about what frightened or worried her. Despite the years we lived together in the same house, I never knew what made her cry in her private moments. Until we really know the whole story, we don't know what burdens a person carries with them from day to day, and what those burdens do to their soul.

For my own good, I knew that I had to forgive my mom. I think that was the hardest part. It took a few years to reach that point, but it was the most important thing I could do if I wanted to live the rest of my life in peace. I have learned that forgiveness doesn't mean you have to fully accept the person who hurt you, or that it means that suddenly things are okay between the two of you. Some things are simply too horrific or the damage is too deep to think that you can have a relationship with the offender. In my mind, forgiveness means letting go of what someone has done to you. Let them deal with the consequences of what they did while you focus instead on letting go of it and moving on with your life. If you can't let go, then long after the offender has forgotten about you, your pain only worsens, affecting your health and the quality of your life. Sometimes this means getting help from a therapist or another professional—a religious leader, perhaps, or just a good friend you can bare your soul to—but you must find a way to let go of the hurt and move on. This is what I had to do, and it helped me in many ways. Each of my brothers and sisters will have to do the same if they hope to find deliverance from what was done to them.

A second piece of advice I would give is to be mindful of your family stories and make a serious effort to record and preserve them for future generations. These stories have a far greater value than most people realize. Sometimes they may not seem like much, but when you hear them in context with other stories, you begin to see a bigger story taking shape. These stories are our legacy for posterity; for better or worse, they will always be a part of us.

As I mentioned earlier in this book, stories are usually created to either share an event or to hide something. This could be an event in one's personal life or one that took place in the context of the larger family. If it's a story to share an event, then the purpose is self-evident. If the story is hiding something, then the challenge is to find out what parts of the story are true and what parts have been created to hide a secret. When you find out what has been made up, you then must decide possible reasons why someone felt the need to make this up, and what were they hiding?

Let me share an example of what I mean. While I was researching some of my own family lines, I was given a photograph that caught my attention. Taken around 1929, this picture included a man in his mid-forties, a woman in her early thirties, and two beautiful young girls, aged around six and four. The smiles on the little girls' faces would normally give the impression that this was a very happy time for them. Yet there was something about it that didn't feel right. There was a look of intense sadness in the eyes of the mother, despite her smile. It immediately made me feel there was a story behind this photograph, and I wanted to know what it was.

I knew the six-year-old girl in the photograph. She was in her late sixties when I first met her. I even had her phone number, since I had been working with her to find all the descendants of this woman's extended family. Her name was Jean. I contacted her and asked if she remembered the photograph. I could hear a slight pause and then a faint sense of pain in her voice as she told me that she did remember the picture. In fact, Jean remembered the photograph very well; it once belonged to her. She had given it to me years ago because it brought her too much pain to see it. She was in her early seventies by the time I asked her about the picture. I tried to be mindful of her pain as I asked her questions. When I shared how I felt about it, she started to cry. I felt bad that I'd caused her tears, but knew I had hit on something. I did my best to console her and said I'd stop asking questions if it was too painful for her to talk about the photograph.

After regaining her composure, she confirmed that there was much more to the picture than one would guess. She said the smiles were not real; they'd only smiled because their mother had asked them to. They knew that their mother was not happy. Their father was an alcoholic and caused great pain to their mother, who wanted so much to have a stable home for her daughters.

Jean told me that her mother had died shortly after this picture was taken. Jean's aunts told her and her little sister that their mother had been doing laundry in the nearby river and slipped on some wet stones. She'd fallen and hit her head on one of the rocks, knocking her unconscious. She fell into the river and drowned, and it took three days for the men in the town to find her body.

I expressed how sorry I was to learn about her loss. I told her that it must have been so hard for her and her sister to grow up without their mother. Jean was silent for a few minutes. In fact, she was quiet for so long that I was tempted to tell her it was okay to stop telling her story if she wanted to. I could tell she was thinking about something more than just her mother falling on the rocks. I knew I had to be quiet and wait until she could sort through her feelings and was ready to talk again.

After a long, awkward pause, Jean told me that the story she'd just recounted wasn't true. It was what the family had shared with everyone in town and with the local newspaper, but it wasn't what actually happened. She then told me the real story of how her mom died.

Jean explained that her mother had married an older man, about fifteen years older than herself. Her dad was a good man, but he was also a hard-core alcoholic. He had tried to quit drinking many times but was never able to beat this terrible habit. Her mother worked a full-time job to help sustain the family because her husband usually spent his earnings drinking and gambling. They had to move so many times that Jean's mother felt Jean and her sister were showing definite signs of trauma and depression.

Jean explained that her mother worked overtime and picked up odd jobs so she could earn extra money. Little by little, her mother was able to hide away enough money to allow her to put a down payment on a house. Her mother was so excited and happy at the thought of buying a home of their own so that her little girls could stay in one place and make friends.

One day, Jean's mother discovered that her husband had found the money she had been saving. Unbeknownst to her, he had spent all of it on drinking and gambling, in one single evening. Her mother was devastated. Her dream of her daughters never having to move again was gone forever.

A few days after this happened, Jean's mother tucked her little girls into bed, read them a story, and told them how much she loved them, and that she was proud of them for being such good girls. She said good night and turned off the lights. Then she went to the river and simply walked into the water and drowned herself. She could no longer bear the pain of seeing her little girls suffer with an alcoholic father, never living in one place long enough to make friends. Her only hope of helping her daughters was now gone, and she couldn't live with that thought. The only part of the original story Jean had shared with me that was true was the fact that it did take three days for the search party to find the body. For Jean, those three days of not knowing what had happened or where her mother was were the worst days of her life.

This real-life story of the death of Jean's mother is an example of how the stories we hear within a family often document an event, while at the same time, they cover up many of the real facts that are simply too embarrassing or too painful for the living family members to deal with. Jean's first story recounts a tragic incident of a mother losing her life doing her daily duties. But the truth was far more traumatic, and a dramatic example of what families must deal with in real life.

Be mindful that stories change as they are passed from person to person. A story told by Grandma may have changed dramatically between the time the experience actually happened and the time it makes its way to a grandchild. That is why it's so important for anyone who hears a story that has been passed down through the generations to pay attention to the details and, where possible, to do some research and document the details of what they find.

We don't like incomplete stories; we prefer order and completeness. Because of this, our minds will find a way to fill in any holes in a story with information, whether fact or fiction. With each new storyteller, new holes appear, with new pieces of information to fill in those holes, leading to more changes and the evolution of a family story. This is especially true when children hear a story and it doesn't make sense to them. They will change details and remember it in a way that makes sense to them.

Upon doing a bit more research, I found that Jean's second version of what happened to her mother was quite accurate. I think Jean learned the truth about her mother's death as an adult, at her father's deathbed.

The reason I shared this story is to underscore the importance of recording all the facts you can when you first hear a story. Do some research and see what details you can find that might help to clarify the event. If it's a story that is known by several family members, try to record everyone's version. You may be surprised by the variations that have crept into the story. Any one version may hold a clue to help bring you closer to the truth. Be mindful that some stories will be heartbreaking; others might possibly demonize a celebrated ancestor; while still others may create an image of someone who is almost too perfect to be true. The facts are always at the mercy of the storyteller.

Like my sister Holly, you may feel that some of your family stories need to die. Some are painful enough that you may feel the pain needs to stop with this generation. While I believe that in a few very rare instances that might be true, I would still be extremely careful about losing any family story. As much as possible, I try not to let any story I've found die. If it's too painful or embarrassing, I keep a copy of it tucked away. Generally, the generations that follow are far more understanding and less likely to be traumatized by the story than those who lived during the time the event took place.

It was the stories I heard as a child that, as an adult, convinced me we needed to create a DNA standard against which my siblings and I could be tested. It was these stories that eventually resulted in each of us learning who our biological fathers were. It led us to find half-siblings and to discover our real genetic legacy. It was our stories that, in the end, gave us a remarkable gift of self-discovery.

A third piece of advice I would give to the reader is to begin now to gather DNA samples of family members. Start with your own family members and then expand from there. Gathering a DNA sample is as simple as taking a cotton swab and rubbing it on the inside of your cheek. Let it dry thoroughly and then put it away where it will be safe from contamination. Technology and medical discoveries are advancing so quickly that preserving DNA samples of older family members will mostly likely prove to be a great asset as you deal with health issues in the future. I won't go into detail here on the dos and don'ts of gathering and preserving DNA samples; there are plenty of good articles online that will describe how to do this. But start gathering them sooner rather than later. Personal experience has shown me that it's harder and more expensive to get information about DNA from a person after they are dead than when they are still living. While that sounds obvious, it may surprise you to learn how many people don't do anything about gathering and preserving DNA samples from family members.

The fourth piece of advice I would like to share is that we all need to understand that there are no perfect families. Ward and June Cleaver were not real people! *Leave It to Beaver* wasn't real. Every family struggles with issues, which in many cases create grief, sadness, stress, and even shame. It's all part of living in a family.

With all families, it can take a lot of time, patience, and forgiveness to overcome the past and learn to live with how things turned out. For some, the damage is so great that there will be no forgiveness from the children. I once heard it said that if we cannot forgive and forget, at least we can forgive and move on. The sooner we let go of the unrealistic idea of the perfect family, the sooner we will be able to accept our own family for what it is, and enjoy the unique experiences and qualities it has to offer. Forgive, let go, and move on.

Always remember that every family is different, and that's really a wonderful thing. We don't get to choose our parents or our siblings. We have absolutely no say in the matter. We would do well to not be so easily offended by the shortcomings of our family members, and to try our best to see the good in those we grew up with. It just makes life a lot easier when we do.

I must say that our parents were unique. In so many ways, they certainly made my life interesting, and I think each of my siblings would agree. We are quite different from each other. We have all faced our share of challenges, and we all have plenty of good qualities that have made the lives of each one of us easier and more rewarding. I don't expect anyone would consider our family typical in any way, but we enjoy our time with each other. We have shared so many experiences that make a family happy, and most importantly, we still love each other.

You can embrace your unique story, or you can discard it and run away from it. But life can be a lot more difficult if you turn your back on your family. We all need and want a place that we can go back to and find acceptance when times are hard. I have learned to embrace my family and be grateful for the experiences that have shaped me.

8

WHERE TO FROM HERE?

As a family, we have come a long way over the past six years in our search for answers. We have learned a lot of new information that has dramatically changed our lives in ways we never could have imagined. Am I finished with my search? No, not yet. I have a few more tasks I would like to take care of before I feel like I can put this aside and call it finished.

First, I would like to see if I can locate any offspring that Mark Anderson might have fathered. While I am convinced that he never fathered any children, I cannot prove that at this point. Given the circumstances, I honestly think he was sterile, but until I can prove otherwise, the possibility exists that there could be sons and daughters fathered by Mark Anderson that we don't know about.

Several well-known DNA testing companies allow customers to import a person's "raw" DNA data into their databases. If I can work with our friend Jack Anderson to get Mark's DNA sample converted into raw DNA data, I can add it to the databases of AncestryDNA, 23andMe, and half a dozen other databases. Once I do that, I can see if we come up with a match with any one of the millions of names in those databases. Even if I don't get a hit right away, I know that each year, millions of new names are added, giving me a continuously new flow of names to check. If, over a five-year period, I find no match, then I will feel comfortable concluding that Mark did not conceive any children.

Second, I eventually want to see if I can make contact with some of our half-siblings. I know that Tim and I would like to see photographs of our biological father and our grandparents. We'd also like to get medical information that we can pass down to our children. Photographs, stories, and insight into each of our fathers' physical descriptions will add a whole new dimension to a relationship. It's one thing to know who we are related to, but knowing something about who these men really were and what they did with their lives could help create a bond that might allow us to feel more at peace about our relationships with them. We could possibly even feel a sense of pride knowing we are a part of our father's genetic line.

Third, I would like to talk with others who have gone through a similar experience in their lives. I doubt my siblings will want to do this, but I definitely do. I've worked through all the phases of grief—denial, anger, bargaining, etc.—that most people go through when dealing with such a discovery, but being part of a community of others who have gone through, or are currently going through, the same type of experience can be helpful. I also want to help others who are just beginning to deal with the shock of discovering their new parent relationship. I feel like I've learned some valuable lessons that could help others navigate their experience without quite as much trauma.

During my search for answers, I have learned that there is a name for people who have discovered that one or both parents are not their biological parents. We are referred to as subjects of a non-paternity event, or an NPE. On occasion, if the maternal side is in question, the acronym is defined as a non-parental (or non-parent) event. In most cases, the term *non-parent event* is used across the board. An NPE can result from:

- An official legal adoption.
- An unofficial non-legal adoption (where a child is "given" to another family member or a friend who has no children).
- An extramarital affair.

- Abandonment.
- A sperm bank donation.
- Rape.
- Kidnapping as an infant.
- Being switched shortly after birth, usually in the hospital.

In some cases, both the mother and a child's non-biological father may be fully aware of the circumstances surrounding a child's conception and birth. However, on occasion, a father may have no idea that the child he is raising is a result of an NPE. If the child was conceived by another man without the knowledge of the current father, then finding out the truth through the use of a DNA test is often just as traumatic for the father (both the non-biological father and, in some cases, the biological father) as it is for the child.

With the availability of inexpensive, easy-to-use home DNA tests, thousands of people are finding surprises they did not expect to see as their DNA test results come back. I have not found specific numbers as to how many people have discovered that they are a result of a non-paternity event, but one online support group called DNA NPE Friends boasts more than 4,700 members (as of January 2019) after being online for about eighteen months. That gives you a fair idea that the number of people who are subjects of an NPE is likely higher than most people realize.

Unfortunately, online support groups are woefully lacking at this time. I found almost nothing of any real value when I did an in-depth online search for a support site. However, I did find one site on Facebook that looks like it's got a lot of great potential. It's a private group called DNA NPE Friends (https://www.facebook.com/groups/NPEGateway). DNA NPE Friends was started by Catherine St Clair and one new online friend who discovered that their birth was the result of an NPE. Like me, St Clair was desperate to find someplace to get help, but found nothing. So, in June of 2017, she started the DNA NPE Friends website with the intention of providing a place for like-minded people to get help dealing with their feelings, and to trade stories and give each other the support they need. I was so impressed with it that I joined the group, and I've found it very helpful.

DNA NPE Friends is a "membership by invitation only" group that involves a gateway vetting process to apply. Applications are reviewed by system administrators to ensure that applicants are serious about being part of this group. Once an application is accepted, the new member is offered the opportunity to join the private section, where the general public is not able to read any of the posts. Members can converse with each other honestly, knowing that comments are protected from outsiders whose only interest is to read sensational stories. The site administrators go out of their way to provide a safe and secure place for members to converse with each other and carry on honest, heartfelt conversations.

Inspired by DNA NPE Friends, in 2018 a nonprofit organization called NPE Friends Fellowship was introduced. The NPE Friends Fellowship provides practical support and resources for the NPE community (www.NPEFellowship.org).

I also found a site called WatershedDNA.com (https://www.watersheddna.com), an online genetic counseling service operated by a woman named Brianne Kirkpatrick. She helps people who are struggling to find answers to their DNA test results. She provides a wide range of personal counseling services for a fee.

I have found that sharing my story with others who have gone through similar experiences is very comforting. It has helped me to see that literally thousands of other people have gone through similar experiences, discovering that their personal history is not what they thought it was. If you find that one or both of your parents is not your biological parent, I strongly recommend that you don't try to go through the emotional turmoil alone. Getting feedback and support from a professional therapist or others who have gone through this experience themselves can provide you with the help and perspective you need when your world is turning upside down.

In so many ways, what we have discovered has started each of us on a journey—toward a new identity, new family members, a new image of who we are (and are not), a new health history, and, most importantly, a new sense of self-worth. We must be kind to ourselves and know that we are by no means alone. With time, we will find that we are part of a much bigger segment of the population than we ever dreamed possible just a few decades ago.

I take some comfort in knowing that my family is not alone on this strange journey. In the meantime, I am going to enjoy watching the looks on people's faces as I tell them about my family story. Somehow, seeing their expressions of shock and surprise takes a lot of the sting out of the reality of what I've learned about my family and who I am. As long as I have been given this new story, I intend to have some fun with it.

FAMILY TREE

Note to the reader: To help you understand the familial relationships in this story, I've included a family tree on the following page. However, I strongly suggest reading the entire story before turning the page to avoid spoilers.

Linda and Mark Anderson with their children. To the right of each child is listed the first name of their biological father. DNA testing for all but Neil has proven that Mark Anderson is not the biological father of any of the nine children below. Neil has not provided a DNA sample to test.

Mother		**Non-Biological Father**
Linda Anderson		Mark Anderson
	↓	
Children		**Biological Father**
1. Holly	→	Sam
2. Judy	→	Sam
3. Gloria	→	Sam
4. Paul	→	Michael (Father's name provided by Linda)
5. Diane	→	John
6. Neil	→	Dennis (Father's name provided by Linda)
7. Steve	→	Timmy
8. Tim	→	Timmy
9. Carlee	→	Peter (Father's name provided by Linda)

NOTES

3. CREATING THE STANDARD

1. Resnick, Richard. "Welcome to the Genomic Revolution." TED Talk, July 2011. Boston, Massachusetts. https://www.ted.com/talks/richard_resnick_welcome_to_the_genomic_revolution
2. Russell, Judy G. "The Ethics of DNA Testing," *The Legal Genealogist*, 2012. https://www.legalgenealogist.com/2012/11/18/the-ethics-of-dna-testing/#respond

4. SURPRISE! DEALING WITH THE RESULTS

1. Kirkpatrick, Brianne, and Jennifer Harstein. Interview with Megyn Kelly, *Megyn Kelly Today*, MSNBC. September 7, 2018.

5. FINDING THE ANSWERS

1. Griffeth, Bill. *The Stranger in My Genes: A Memoir*. Boston: New England Historic Genealogical Society, 2016.

7. QUESTIONS AND ANSWERS

1. Kirkpatrick, Brianne, and Jennifer Harstein. Interview with Megyn Kelly, *Megyn Kelly Today*, MSNBC. September 7, 2018.

2. Hamer, Ashley. "Remembering the Same Event Differently? That's the Rashomon Effect." Podcast audio, September 9, 2016. https://curiosity.com/topics/the-rashomon-effect-refers-to-conflicting-accounts-of-the-same-event-curiosity/.

3. Kirkpatrick, Brianne, and Jennifer Harstein, Interview with Megyn Kelly, *Meygn Kelly Today*, MSNBC. September 7, 2018.

4. Ibid.

ABOUT THE AUTHOR

Stephen F. Anderson retired in 2016 after working for more than twenty-six years for FamilySearch International, one of the largest online family history sites in the world. During the last five years of his work with FamilySearch, he moved to the newly formed marketing division to work as the FamilySearch blogger, focusing on a large variety of family history topics, including how to use genetic tools to expand the story of one's family.

Steve has given presentations at family history conferences for more than fifteen years, focusing on using DNA tests and other genetic tools to do family history research, including a case study presentation about his own family story. He has also presented on the ethics of using DNA as a family history research tool. He has been invited on two different occasions to present his story on the syndicated radio show and podcast *Extreme Genes*, hosted by Scott Fisher. Well-known writer A. J. Jacobs interviewed Steve and used his story in his latest book, *It's All Relative* (chapter 20).

Steve has published a collection of oral histories for the Lindon City Historical Society in Utah. He's also written articles for various national and international magazines on how to do family history research and oral histories, as well as more than three hundred articles for the FamilySearch blog.

Ingram Content Group UK Ltd.
Milton Keynes UK
UKHW011303130723
425080UK00022B/206